**Lectures in Mathematics
ETH Zürich**
Department of Mathematics
Research Institute of Mathematics

Managing Editor:
Michael Struwe

# Jürgen Moser
# Selected Chapters in the Calculus of Variations

Lecture Notes by Oliver Knill

Springer Basel AG

Author:
Jürgen Moser †
Department of Mathematics, ETH Zürich, Switzerland

Contact address:
Oliver Knill
Department of Mathematics
Harvard University
Cambridge, MA 02138
USA
e-mail: knill@math.harvard.edu

2000 Mathematical Subject Classification 49Q20, 53D25

A CIP catalogue record for this book is available from the
Library of Congress, Washington D.C., USA

Bibliographic information published by Die Deutsche Bibliothek
Die Deutsche Bibliothek lists this publication in the Deutsche Nationalbibliografie; detailed
bibliographic data is available in the Internet at <http://dnb.ddb.de>.

ISBN 978-3-7643-2185-7     ISBN 978-3-0348-8057-2 (eBook)
DOI 10.1007/978-3-0348-8057-2

© 2003 Springer Basel AG
Originally published by Birkhäuser Verlag, Basel, Switzerland in 2003
Printed on acid-free paper produced from chlorine-free pulp. TCF

ISBN 978-3-7643-2185-7

9 8 7 6 5 4 3 2 1                     www.birkhauser-science.com

# Contents

# 0.1 Introduction

These lecture notes describe a new development in the calculus of variations which is called **Aubry–Mather–Theory**.

The starting point for the theoretical physicist Aubry was a model for the description of the motion of electrons in a two-dimensional crystal. Aubry investigated a related discrete variational problem and the corresponding minimal solutions.

On the other hand, Mather started with a specific class of area-preserving annulus mappings, the so-called **monotone twist maps**. These maps appear in mechanics as Poincaré maps. Such maps were studied by Birkhoff during the 1920s in several papers. In 1982, Mather succeeded to make essential progress in this field and to prove the existence of a class of closed invariant subsets which are now called **Mather sets**. His existence theorem is based again on a variational principle.

Although these two investigations have different motivations, they are closely related and have the same mathematical foundation. We will not follow those approaches but will make a connection to classical results of Jacobi, Legendre, Weierstrass and others from the 19th century.

Therefore in Chapter I, we will put together the results of the classical theory which are the most important for us. The notion of **extremal fields** will be most relevant.

In Chapter II we will investigate variational problems on the 2-dimensional torus. We will look at the corresponding global minimals as well as at the relation between minimals and extremal fields. In this way, we will be led to Mather sets.

Finally, in Chapter III, we will learn the connection with monotone twist maps, the starting point for Mather's theory. In this way we will arrive at a discrete variational problem which forms the basis for Aubry's investigations.

This theory has additional interesting applications in differential geometry. One of those is the geodesic flow on two-dimensional surfaces, especially on the torus. In this context the **minimal geodesics** play a distinguished role. They were investigated by Morse and Hedlund in 1932.

As Bangert has shown, the theories of Aubry and Mather lead to new results for the geodesic flow on the two-dimensional torus. As the last section of these lecture notes will show, the restriction to two dimensions is essential. These differential geometric questions are treated at the end of the third chapter.

The beautiful survey article of Bangert should be at hand when reading these lecture notes.

Our description aims less at generality. We rather aim to show the relations of newer developments with classical notions like extremal fields. Mather sets will appear as 'generalized extremal fields' in this terminology.

For the production of these lecture notes I was assisted by O. Knill to whom I want to express my thanks.

Zürich, September 1988, J. Moser

## 0.2   On these lecture notes

These lectures were presented by J. Moser in the spring of 1988 at the Eidgenössische Technische Hochschule (ETH) Zürich. Most of the students were enrolled in the 6th to the 8th semester of the 4 year Mathematics curriculum. There were also graduate students and visitors from the research institute at the ETH (FIM) in the auditorium.

In the last decade, the research on this particular topic of the calculus of variations has made some progress. A few hints to the literature are listed in an Appendix. Because some important questions are still open, these lecture notes are maybe of more than historical value.

The notes were typed in the summer of 1988. J. Moser had looked carefully through the notes in September 1988. Because the text editor in which the lecture were originally written is now obsolete, the typesetting was done from scratch with LATEX in the year 2000. The original had not been changed except for small, mostly stylistic or typographical corrections. In 2002, an English translation was finished and figures were added.

Cambridge, MA, December 2002, O. Knill

# Chapter 1

# One-dimensional variational problems

## 1.1 Regularity of the minimals

Let $\Omega$ be an open region in $\mathbb{R}^{n+1}$. We assume that $\Omega$ is simply connected. A point in $\Omega$ has the coordinates $(t, x_1, ..., x_n) = (t, x)$. Let $F = F(t, x, p) \in C^r(\Omega \times \mathbb{R}^n)$ with $r \geq 2$ and let $(t_1, a)$ and $(t_2, b)$ be two points in $\Omega$. The space

$$\Gamma := \{ \gamma : t \to x(t) \in \Omega \mid x \in C^1[t_1, t_2], \; x(t_1) = a, x(t_2) = b \}$$

consists of all continuously differentiable curves which start at $(t_1, a)$ and end at $(t_2, b)$. On $\Gamma$ is defined the functional

$$I(\gamma) = \int_{t_1}^{t_2} F(t, x(t), \dot{x}(t)) \, dt .$$

**Definition**. We say that $\gamma^* \in \Gamma$ is **minimal** in $\Gamma$ if

$$I(\gamma) \geq I(\gamma^*), \; \forall \gamma \in \Gamma .$$

We first search for necessary conditions for a minimum of $I$ while assuming the existence of a minimal.

**Remark**. A minimum does not need to exist in general:

- It is possible that $\Gamma = \emptyset$.

- It is also possible that a minimal $\gamma^*$ is contained only in $\overline{\Omega}$.

- Finally, the infimum could exist without the minimum being achieved.

**Example.** Let $n = 1$ and $F(t, x, \dot{x}) = t^2 \cdot \dot{x}^2, (t_1, a) = (0, 0), (t_2, b) = (1, 1)$. We have

$$\gamma_m(t) = t^m, \ I(\gamma_m) = \frac{1}{m + 3}, \ \inf_{m \in \mathbb{N}} I(\gamma_m) = 0,$$

but for all $\gamma \in \Gamma$ one has $I(\gamma) > 0$.

---

**Theorem 1.1.1.** *If $\gamma^*$ is minimal in $\Gamma$, then*

$$F_{p_j}(t, x^*, \dot{x}^*) = \int_{t_1}^{t} F_{x_j}(s, x^*, \dot{x}^*) \, ds = const$$

*for all $t_1 \leq t \leq t_2$ and $j = 1, ..., n$. These equations are called* **integrated Euler equations.**

---

**Definition.** One calls $\gamma^*$ **regular** if $\det(F_{p_i p_j}) \neq 0$ for $x = x^*, p = \dot{x}^*$.

---

**Theorem 1.1.2.** *If $\gamma^*$ is a regular minimal, then $x^* \in C^2[t_1, t_2]$ and one has for $j = 1, \ldots, n$,*

$$\frac{d}{dt} F_{p_j}(t, x^*, \dot{x}^*) = F_{x_j}(t, x^*, \dot{x}^*) \tag{1.1}$$

*These equations are called* **Euler equations.**

---

**Definition.** An element $\gamma^* \in \Gamma$ satisfying the Euler equations (1.1) is called an **extremal** in $\Gamma$.

**Warning.** Not every extremal solution is a minimal!

*Proof of Theorem* 1.1.1. We assume that $\gamma^*$ is minimal in $\Gamma$. Let $\xi \in C_0^1(t_1, t_2) = \{x \in C^1[t_1, t_2] \mid x(t_1) = x(t_2) = 0 \}$ and $\gamma_\epsilon : t \mapsto x(t) + \epsilon \xi(t)$. Because $\Omega$ is open and $\gamma \in \Omega$, also $\gamma_\epsilon \in \Omega$ for small enough $\epsilon$. Therefore,

$$\begin{aligned}
0 &= \frac{d}{d\epsilon} I(\gamma_\epsilon)|_{\epsilon=0} \\
&= \int_{t_1}^{t_2} \sum_{j=1}^{n} \left( F_{p_j}(s)\dot{\xi}_j + F_{x_j}(s)\xi_j \right) ds \\
&= \int_{t_1}^{t_2} (\lambda(t), \dot{\xi}(t)) \, dt
\end{aligned}$$

with $\lambda_j(t) = F_{p_j}(t) - \int_{t_1}^{t_2} F_{x_j}(s) \, ds$. Theorem 1.1.1 is now a consequence of the following lemma. $\qquad \square$

**Lemma 1.1.3.** *If* $\lambda \in C[t_1, t_2]$ *and*

$$\int_{t_1}^{t_2} (\lambda, \dot{\xi}) \, dt = 0, \quad \forall \xi \in C_0^1[t_1, t_2]$$

*then* $\lambda = const.$

*Proof.* Define $c = (t_2 - t_1)^{-1} \int_{t_1}^{t_2} \lambda(t) \, dt$ and put $\xi(t) = \int_{t_1}^{t} (\lambda(s) - c) \, ds$. Now $\xi \in C_0^1[t_1, t_2]$. By assumption we have:

$$0 = \int_{t_1}^{t_2} (\lambda, \dot{\xi}) \, dt = \int_{t_1}^{t_2} (\lambda, (\lambda - c)) \, dt = \int_{t_1}^{t_2} (\lambda - c)^2 \, dt \, ,$$

where the last equation followed from $\int_{t_1}^{t_2} (\lambda - c) \, dt = 0$. Because $\lambda$ is continuous, this implies with $\int_{t_1}^{t_2} (\lambda - c)^2 \, dt = 0$ the claim $\lambda = const.$ $\square$

*Proof of Theorem 1.1.2.* Put $y_j^* = F_{p_j}(t, x^*, p^*)$. Since by assumption $\det(F_{p_i p_j}) \neq 0$ at every point $(t, x^*(t), \dot{x}^*(t))$, the implicit function theorem assures that functions $p_k^* = \phi_k(t, x^*, y^*)$ exist, which are locally $C^1$. From Theorem 1.1.1 we know

$$y_j^* = const - \int_{t_1}^{t} F_{x_j}(s, x^*, \dot{x}^*) \, ds \in C^1 \tag{1.2}$$

and so

$$\dot{x}_k^* = \phi_k(t, x^*, y^*) \in C^1 \, .$$

Therefore $x_k^* \in C^2$. The Euler equations are obtained from the integrated Euler equations in Theorem 1.1.1. $\square$

**Theorem 1.1.4.** *If* $\gamma^*$ *is minimal, then*

$$(F_{pp}(t, x^*, y^*)\zeta, \zeta) = \sum_{i,j=1}^{n} F_{p_i p_j}(t, x^*, y^*)\zeta_i \zeta_j \geq 0$$

*holds for all* $t_1 < t < t_2$ *and all* $\zeta \in \mathbb{R}^n$.

*Proof.* Let $\gamma_\epsilon$ be defined as in the proof of Theorem 1.1.1. Then $\gamma_\epsilon : t \mapsto x^*(t) + \epsilon \xi(t), \xi \in C_0^1$.

$$0 \leq II \quad := \quad \frac{d^2}{(d\epsilon)^2} I(\gamma_\epsilon)|_{\epsilon=0} \tag{1.3}$$

$$= \quad \int_{t_1}^{t_2} (F_{pp}\dot{\xi}, \dot{\xi}) + 2(F_{px}\dot{\xi}, \xi) + (F_{xx}\xi, \xi) \, dt \, . \tag{1.4}$$

$II$ is called the **second variation** of the functional $I$. Let $t \in (t_1, t_2)$ be arbitrary. We construct now special functions $\xi_j \in C_0^1(t_1, t_2)$:

$$\xi_j(t) = \zeta_j \psi(\frac{t - \tau}{\epsilon}) \,,$$

where $\zeta_j \in \mathbb{R}$ and $\psi \in C^1(\mathbb{R})$ by assumption, $\psi(\lambda) = 0$ for $|\lambda| > 1$ and $\int_{\mathbb{R}} (\psi')^2 \, d\lambda = 1$. Here $\psi'$ denotes the derivative with respect to the new time variable $\tau$, which is related to $t$ as follows:

$$t = \tau + \epsilon\lambda, \ \epsilon^{-1}dt = d\lambda \,.$$

The equations

$$\dot{\xi}_j(t) = \epsilon^{-1} \zeta_j \psi'(\frac{t - \tau}{\epsilon})$$

and (1.3) give

$$0 \leq \epsilon^3 II = \int_{\mathbb{R}} (F_{pp}\zeta, \zeta)(\psi')^2(\lambda) \, d\lambda + O(\epsilon) \,.$$

For $\epsilon > 0$ and $\epsilon \to 0$ this means that

$$(F_{pp}(t, x(t), \dot{x}(t))\zeta, \zeta) \geq 0 \,. \qquad \square$$

**Definition.** We call the function $F$ **autonomous**, if $F$ is independent of $t$.

---

**Theorem 1.1.5.** *If $F$ is autonomous, every regular extremal solution satisfies*

$$H = -F + \sum_{j=1}^{n} p_j F_{p_j} = const \,.$$

*The function $H$ is also called the **energy**. In the autonomous case we have therefore energy conservation.*

---

*Proof.* Because the partial derivative $H_t$ vanishes, one has

$$
\begin{aligned}
\frac{d}{dt} H &= \frac{d}{dt}(-F + \sum_{j=1}^{n} p_j F_{p_j}) \\
&= \sum_{j=1}^{n} (-F_{x_j}\dot{x}_j - F_{p_j}\ddot{x}_j + \ddot{x}_j F_{p_j} + \dot{x}_j \frac{d}{dt} F_{p_j}) \\
&= \sum_{j=1}^{n} -F_{x_j}\dot{x}_j - F_{p_j}\ddot{x}_j + \ddot{x}_j F_{p_j} + \dot{x}_j F_{x_j} = 0 \,.
\end{aligned}
$$

Because the extremal solution was assumed to be regular, we could use the Euler equations (Theorem 1.1.2) in the last step. $\qquad \square$

In order to obtain sharper regularity results we change the variational space. We have seen that if $F_{pp}$ is not degenerate, then $\gamma^* \in \Gamma$ is two times differentiable even though the elements in $\Gamma$ are only $C^1$. This was the statement of the regularity Theorem 1.1.2.

We consider now a bigger class of curves

$$\Lambda = \{\gamma : [t_1, t_2] \to \Omega, \ t \mapsto x(t), x \in \mathrm{Lip}[t_1, t_2], x(t_1) = a, x(t_2) = b \} \ .$$

$\mathrm{Lip}[t_1, t_2]$ denotes the space of Lipschitz continuous functions on the interval $[t_1, t_2]$. Note that $\dot{x}$ is now only measurable and bounded. Nevertheless there are results analogous to Theorem 1.1.1 or Theorem 1.1.2:

---

**Theorem 1.1.6.** *If $\gamma^*$ is a minimal in $\Lambda$, then*

$$F_{p_j}(t, x^*, \dot{x}^*) - \int_{t_1}^{t_2} F_{x_j}(s, x^*, \dot{x}^*) \, ds = const \qquad (1.5)$$

*for Lebesgue almost all $t \in [t_1, t_2]$ and all $j = 1, ..., n$.*

---

*Proof.* As in the proof of Theorem 1.1.1 we put $\gamma_\epsilon = \gamma + \epsilon \xi$, but this time, $\xi$ is in

$$\mathrm{Lip}_0[t_1, t_2] := \{\gamma : t \mapsto x(t) \in \Omega, \ x \in \mathrm{Lip}[t_1, t_2], x(t_1) = x(t_2) = 0 \} \ .$$

So,

$$
\begin{aligned}
0 &= \frac{d}{d\epsilon} I(\gamma_\epsilon)|_{\epsilon=0} \\
&= \lim_{\epsilon \to 0} (I(\gamma_\epsilon) - I(\gamma_0))/\epsilon \\
&= \lim_{\epsilon \to 0} \int_{t_1}^{t_2} [F(t, \gamma^* + \epsilon\xi, \dot{\gamma}^* + \epsilon\dot{\xi}) - F(t, \gamma^*, \dot{\gamma}^*)]/\epsilon \, dt \ .
\end{aligned}
$$

To take the limit $\epsilon \to 0$ inside the integral, we use Lebesgue's dominated convergence theorem: for fixed $t$ we have

$$\lim_{\epsilon \to 0} [F(t, \gamma^* + \epsilon\xi, \dot{\gamma}^* + \epsilon\dot{\xi}) - F(t, \gamma^*, \dot{\gamma}^*)]/\epsilon = (F_x, \xi) + (F_p, \dot{\xi})$$

and

$$\frac{F(t, \gamma^* + \epsilon\xi, \dot{\gamma}^* + \epsilon\dot{\xi}) - F(t, \gamma, \dot{\gamma})}{\epsilon} \le \sup_{s \in [t_1, t_2]} |F_x(s, x(s), \dot{x}(s)| \xi(s) + |F_p(s, x(s)| \dot{\xi}(s) \ .$$

The last expression is in $L^1[t_1, t_2]$. Applying Lebesgue's theorem gives

$$0 = \frac{d}{d\epsilon} I(\gamma_\epsilon)|_{\epsilon=0} = \int_{t_1}^{t_2} (F_x, \xi) + (F_p, \dot{\xi}) \, dt = \int_{t_1}^{t_2} \lambda(t)\dot{\xi} \, dt$$

with $\lambda(t) = F_p - \int_{t_1}^{t_2} F_x \, ds$. This is bounded and measurable.

Define $c = (t_2 - t_1)^{-1} \int_{t_1}^{t} \lambda(t)\ dt$ and put $\xi(t) = \int_{t_1}^{t_2}(\lambda(s) - c)\ ds$. We get $\xi \in \text{Lip}_0[t_1, t_2]$ and in the same way as in the proof of Theorem 1.1.4 or Lemma 1.1.3 one concludes

$$0 = \int_{t_1}^{t_2} (\lambda, \dot{\xi})\ dt = \int_{t_1}^{t_2} (\lambda, (\lambda(t) - c)))\ dt = \int_{t_1}^{t_2} (\lambda - c)^2\ dt\ ,$$

where the last equation followed from $\int_{t_1}^{t_2}(\lambda - c)\ dt = 0$. This means that $\lambda = c$ for almost all $t \in [t_1, t_2]$.                                                               $\square$

---

**Theorem 1.1.7.** *If $\gamma^*$ is a minimal in $\Lambda$ and $F_{pp}(t, x, p)$ is positive definite for all $(t, x, p) \in \Omega \times \mathbb{R}^n$, then $x^* \in C^2[t_1, t_2]$ and*

$$\frac{d}{dt} F_{p_j}(t, x^*, \dot{x}^*) = F_{x_j}(t, x^*, \dot{x}^*)$$

*for $j = 1, ..., n$.*

---

*Proof.* The proof uses the integrated Euler equations in Theorem 1.1.1. It makes use of the fact that a solution of the implicit equation $y = F_p(t, x, p)$ for $p = \Phi(t, x, y)$ is **globally unique**. Indeed: if two solutions $p$ and $q$ would exist with

$$y = F_p(t, x, p) = F_q(t, x, q)\ ,$$

it would imply that

$$0 = (F_p(t, x, p) - F_p(t, x, q), p - q) = (A(p - q), p - q)$$

with

$$A = \int_0^1 F_{pp}(t, x, p + \lambda(q - p))\ d\lambda$$

and because $A$ was assumed to be positive definite, $p = q$ follows.

From the integrated Euler equations we know that

$$y(t) = F_p(t, x, \dot{x})$$

is continuous with bounded derivatives. Therefore $\dot{x} = \Phi(t, x, y)$ is absolutely continuous. Integration leads to $x \in C^1$. The integrable Euler equations of Theorem 1.1.1 tell now that $F_p$ is even in $C^1$ and we get, with the already proven global uniqueness result, that $\dot{x}$ is in $C^1$ and hence that $x$ is in $C^2$. We obtain the Euler equations by differentiating (1.5).                                    $\square$

A remark on newer developments: we have seen that a minimal $\gamma^* \in \Lambda$ is two times continuously differentiable. A natural question is whether we obtain such smooth minimals also in bigger variational spaces as in

$$\Lambda_a = \{\gamma : [t_1, t_2] \to \Omega, t \mapsto x(t), \ x \in W^{1,1}[t_1, t_2], x(t_1) = a, x(t_2) = b\} ,$$

the space of **absolutely continuous** curves $\gamma$. One has in that case to deal with singularities for minimal $\gamma$ which form a set of measure zero. Also, the infimum in this class $\Lambda_a$ can be smaller than the infimum in the Lipschitz class $\Lambda$. This is called the **Lavremtiev phenomenon**. Examples of this kind have been given by Ball and Mizel. One can read more about it in the work of Davie [9].

In the next chapter we consider the special case when $\Omega = \mathbb{T}^2 \times \mathbb{R}$. We will also work in the bigger function space

$$\Xi = \{\gamma : [t_1, t_2] \to \Omega, t \to x(t), x \in W^{1,2}[t_1, t_2], x(t_1) = a, x(t_2) = b\} ,$$

and assume some growth conditions on $F = F(t, x, p)$ for $p \to \infty$.

## 1.2  Examples

### Example 1) Free motion of a mass point on a manifold.

Let $M$ be an $n$-dimensional Riemannian manifold with metric $g_{ij} \in C^2(M)$, (where the matrix-valued function $g_{ij}$ is of course assumed to be symmetric and positive definite). Let

$$F(x, p) = \frac{1}{2} g_{ij}(x) p^i p^j .$$

We use the **Einstein summation convention**, which tells us to sum over lower and upper indices.

On the manifold $M$ two points $a$ and $b$ in the same chart $U \subset M$ are given. $U$ is homeomorphic to an open region in $\mathbb{R}^n$ and we define $W = U \times \mathbb{R}$. We also fix two time parameters $t_1$ and $t_2$ in $\mathbb{R}$. The space $\Lambda$ can now be defined as above. From Theorem 1.1.2 we know that a minimal $\gamma^*$ to

$$I(x) = \int_{t_1}^{t_2} F(t, x, x) dt = \int_{t_1}^{t_2} g_{ij}(x) \dot{x}^i \dot{x}^j \, dt \tag{1.6}$$

has to satisfy the Euler equations

$$F_{p_k} = g_{ki} p^i ,$$
$$F_{x_k} = \frac{1}{2} \frac{\partial}{\partial x^k} g_{ij}(x) p^i p^j .$$

The Euler equations for $\gamma^*$ can, using the identity

$$\frac{1}{2}\frac{\partial}{\partial x^j}g_{ik}(x)\dot{x}^i\dot{x}^j = \frac{1}{2}\frac{\partial}{\partial x^i}g_{jk}(x)\dot{x}^i\dot{x}^j$$

and the **Christoffel symbols**

$$\Gamma_{ijk} = \frac{1}{2}\left[\frac{\partial}{\partial x^i}g_{jk}(x) + \frac{\partial}{\partial x^j}g_{ik}(x) - \frac{\partial}{\partial x^k}g_{ij}(x)\right],$$

be written as

$$g_{ki}\ddot{x}^i = -\Gamma_{ijk}\dot{x}^i\dot{x}^j,$$

which are with

$$g^{ij} := g_{ij}^{-1}, \ \Gamma_{ij}^k := g^{lk}\Gamma_{ijl}$$

of the form

$$\ddot{x}^k = -\Gamma_{ij}^k\dot{x}^i\dot{x}^j.$$

These are the differential equations describing **geodesics**. Since $F$ is independent of $t$, it follows from Theorem 1.1.5 that

$$p^k F_{p^k} - F = p^k g_{ki}p^i - F = 2F - F = F$$

are constant along the orbit. This can be interpreted as the **kinetic energy**. The Euler equations describe the orbit of a mass point in $M$ which moves from $a$ to $b$ under no influence of any exterior forces.

**Example 2) Geodesics on a manifold.**

Using the notation of the last example we consider the new function

$$G(t,x,p) = \sqrt{g_{ij}(x)p^ip^j} = \sqrt{2F}.$$

The functional

$$I(\gamma) = \int_{t_1}^{t_2}\sqrt{g_{ij}(x)\dot{x}^i\dot{x}^j}\,dt$$

gives the **arc length** of $\gamma$. The Euler equations

$$\frac{d}{dt}G_{p^i} = G_{x^i} \tag{1.7}$$

can, using the previous function $F$, be written as

$$\frac{d}{dt}\frac{F_{p^i}}{\sqrt{2F}} = \frac{F_{x^i}}{\sqrt{2F}} \tag{1.8}$$

and these equations are satisfied if

$$\frac{d}{dt}F_{p^i} = F_{x^i} \tag{1.9}$$

because $\frac{d}{dt}F = 0$. So we obtain the same equations as in the first example. Equations (1.8) and (1.9) are however not equivalent because a reparameterization of time $t \mapsto \tau(t)$ leaves only equation (1.8) invariant and not equation (1.9). The distinguished parameterization for the extremal solution of (1.9) is proportional to the arc length.

The relation of the two variational problems which we have met in examples 1) and 2) is a special case of the **Maupertius principle** which we mention here for completeness:

Let the function $F$ be given by

$$F = F_2 + F_1 + F_0 \; ,$$

where $F_i$ are independent of $t$ and **homogeneous** of degree $j$. ($F_j$ is homogeneous of degree $j$, if $F_j(t, x, \lambda p) = \lambda F_j(t, x, p)$ for all $\lambda \in \mathbb{R}$.) The term $F_2$ is assumed to be positive definite. Then the energy

$$pF_p - F = F_2 - F_0$$

is invariant. We can assume without loss of generality that we are on an energy surface $F_2 - F_0 = 0$. With $F_2 = F_0$, we get

$$F = F - (\sqrt{F_2} - \sqrt{F_0})^2 = 2\sqrt{F_2 F_0} - F_1 = G$$

and

$$I(x) = \int_{t_1}^{t_2} G \, dt = \int_{t_1}^{t_2} (2\sqrt{F_2 F_0} - F_1) \, dt$$

is independent of the parameterization. Therefore the right-hand side is homogeneous of degree 1. If $x$ satisfies the Euler equations for $F$ and the energy satisfies $F_2 - F_1 = 0$, then $x$ satisfies also the Euler equations for $G$. The case derived in examples 1) and 2) correspond to $F_1 = 0$, $F_0 = c > 0$.

---

**Theorem 1.2.1.** *(Maupertius principle) If $F = F_2 + F_1 + F_0$, where $F_j$ are homogeneous of degree $j$ and independent of $t$ and $F_2$ is positive definite, then every $x$ on the energy surface $F_2 - F_0 = 0$ satisfies the Euler equations*

$$\frac{d}{dt} F_p = F_x$$

*with $F_2 = F_0$ if and only if $x$ satisfies the Euler equations $\frac{d}{dt} G_p = G_x$.*

---

*Proof.* If $x$ is a solution of $\frac{d}{dt} F_p = F_x$ with $F_2 - F_0 = 0$, then

$$\delta \int G \, dt = \delta \int F \, dt - 2 \int (\sqrt{F_2} - \sqrt{F_0}))\delta(\sqrt{F_2} - \sqrt{F_0}) = 0 \; .$$

(Here $\delta I$ denotes the first variation of the functional $I$.) Therefore $x$ is a critical point of $\int G \, dt = \int (2\sqrt{F_2 F_0} - F_1) \, dt$ and $x$ satisfies the Euler equations $\frac{d}{dt} G_p = G_x$. On the other hand, if $x$ is a solution of the Euler equations for $G$, we reparameterize $x$ in such a way that, with the new time

$$ t = t(s) = \int_{t_1}^{s} \frac{\sqrt{F_2(\tau, x(\tau), \dot{x}(\tau))}}{\sqrt{F_0(\tau, x(\tau), \dot{x}(\tau))}} \, d\tau \; , $$

$x(t)$ satisfies the Euler equations for $F$, if $x(s)$ satisfies the Euler equations for $G$. If $x(t)$ is on the energy surface $F_2 = F_0$, then $x(t) = x(s)$ and $x$ satisfies also the Euler equations for $F$.                                                                          $\square$

We see from Theorem 1.2.1 that in the case $F_1 = 0$, the extremal solutions of $F$ even correspond to the geodesics in the Riemannian metric

$$ g_{ij}(x) p^i p^j = (p, p)_x = 4 F_0(x, p) F_2(x, p) \; . $$

This metric $g$ is called the **Jacobi metric**.

**Example 3) A particle in a potential in Euclidean space.**

We consider now the path $x(t)$ of a particle with mass $m$ in Euclidean space $\mathbb{R}^n$, where the particle moves under the influence of a force defined by the potential $U(x)$. An extremal solution to the Lagrange function

$$ F(t, x, p) = mp^2/2 + E - U(x) $$

leads to the Euler equations

$$ m\ddot{x} = -\frac{\partial U}{\partial x} \; . $$

$E$ is then the constant **energy**

$$ E = pF_p - F = mp^2/2 + U \; . $$

The expression $F_2 = mp^2/2$ is positive definite and homogeneous of degree 2. Furthermore $F_0 = E - U(x)$ is homogeneous of degree 0 and $F = F_2 + F_0$. From Theorem 1.2.1 we conclude that the extremal solutions of $F$ with energy $E$ correspond to geodesics of the Jacobi metric

$$ g_{ij}(x) = 2(E - U(x)) \delta_{ij} \; . $$

It is well known that the solutions are not always minimals of the functional. They are stationary solutions in general.

Consider for example the linear pendulum, where the potential is $U(x) = \omega^2 x$ and where we want to minimize

$$I(x) = \int_0^T F(t, x, \dot{x}) \, dt = \int_0^T (\dot{x}^2 - \omega^2 x^2) \, dt$$

in the class of functions satisfying $x(0) = 0$ and $x(T) = 0$. The solution $x \equiv 0$ is a solution of the Euler equations. It is however only a minimal solution if $0 < T \leq \pi/\omega$. (Exercise). If $T > \pi/w$, we have $I(\xi) < I(0)$ for a certain $\xi \in C(0, T)$ with $\xi(0) = \xi(T) = 0$.

**Example 4) Geodesics on the rotationally symmetric torus in $\mathbb{R}^3$**

The rotationally symmetric torus, embedded in $\mathbb{R}^3$, is parameterized by

$$(u, v) \mapsto ((a + b\cos(2\pi v))\cos(2\pi u), (a + b\cos(2\pi v))\sin(2\pi u), b\sin(2\pi v)) \, ,$$

where $0 < b < a$. The metric $g_{ij}$ on the torus is given by

$$
\begin{aligned}
g_{11} &= 4\pi^2(a + b\cos(2\pi v))^2 = 4\pi^2 r^2 \, , \\
g_{22} &= 4\pi^2 b^2 \, , \\
g_{12} &= g_{21} = 0 \, ,
\end{aligned}
$$

so that the line element $ds$ has the form

$$ds^2 = 4\pi^2[(a + b\cos(2\pi v))^2 \, du^2 + b^2 dv^2] = 4\pi^2(r^2 du^2 + b^2 dv^2) \, .$$

Evidently, $v \equiv 0$ and $v \equiv 1/2$ are geodesics, where $v \equiv 1/2$ is a minimal geodesic. The curve $v = 0$ is however not a minimal geodesic!

If $u$ is the time parameter, we can reduce the problem of finding extremal solutions to the functional

$$4\pi^2 \int_{t_1}^{t_2} (a + b\cos(2\pi v))^2 \dot{u}^2 + b^2 \dot{v}^2 \, dt$$

to the question of finding extremal solutions to the functional

$$4\pi^2 b^2 \int_{u_2}^{u_1} F(v, v') \, du \, ,$$

with $u_j = u(t_j)$, where

$$F(v, v') = \sqrt{(\frac{a}{b} + \cos(2\pi v))^2 + (v')^2} = \sqrt{\frac{r^2}{b^2} + (v')^2}$$

with $v' = \frac{dv}{du}$. This worked because our original Lagrange function is independent of $u$. With E. Nöther's theorem we obtain immediately the **angular momentum** as an invariant. This is a consequence of the rotational symmetry of the torus. With $u$ as time, this is a conserved quantity. All solutions are regular and the Euler equations are

$$\frac{d}{du}\left(\frac{v'}{F}\right) = F_v \ .$$

Because $F$ is autonomous, $\frac{dF}{du} = 0$, Theorem 1.1.5 implies energy conservation

$$
\begin{aligned}
E &= v'F_{v'} - F \\
&= \frac{v'^2}{F} - F \\
&= -b^2 r^2/F \\
&= -b^2 r \sin(\psi) = const. \ ,
\end{aligned}
$$

where $r = a + b\cos(2\pi v)$ is the distance to the axes of rotation and where $\sin(\psi) = r/F$. The geometric interpretation is that $\psi$ is the angle between the tangent of the geodesic and the meridian $u = const$. For $E = 0$ we get $\psi = 0 \pmod{\pi}$: the meridians are geodesics. The conserved quantity $r\sin(\psi)$ is called the **Clairaut integral**. It appears naturally as an invariant for a surface of revolution.

**Example 5) Billiards**

To motivate the definition of billiards later on, we first consider the geodesic flow on a two-dimensional smooth Riemannian manifold $M$ homeomorphic to a sphere. We assume that $M$ has a strictly convex boundary in $R^3$. The images of $M$ under the maps

$$z_n : \mathbb{R}^3 \to \mathbb{R}^3, \ (x, y, z) \mapsto (x, y, z/n)$$

$M_n = z_n(M)$ are again Riemannian manifolds. They have the same properties as $M$ and especially possess well-defined geodesic flows. For larger and larger $n$, the manifolds $M_n$ become flatter and flatter. In the limit $n \to \infty$, we end up with a strictly convex and flat region on which the geodesics are straight lines leaving the boundary with the same angle as the impact angle. The limiting system is called **billiards**. If we follow a degenerate geodesic and the successive impact points at the boundary, we obtain a map. This map can be defined also without preliminaries:

Let $\Gamma$ be a convex, smooth and closed curve in the
plane of arc length 1. We fix a point $O$ and an orientation on $\Gamma$. Every point $P$ on $\Gamma$ is now assigned a
real number $s$, the arc-length of the arc from $O$ to $P$
in the positive direction. Let $t$ be the angle between
the line passing through $P$ and the tangent of $\Gamma$ in
$P$. For $t \in (0, \pi)$ this line has a second intersection $P$
with $\Gamma$. To that intersection we can assign two numbers, $s_1$ and $t_1$. If $t = 0$ we put $(s_1, t_1) = (s, t)$ and
for $t = \pi$ we define $(s_1, t_1) = (s + 1, t)$.

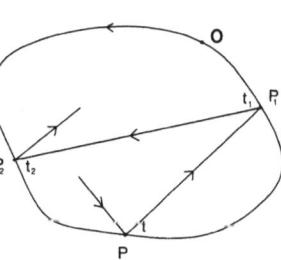

Let $\phi$ be the map $(s, t) \mapsto (s_1, t_1)$. It is a map from the closed annulus

$$A = \{(s, t) \mid s \in \mathbb{R}/\mathbb{Z}, t \in [0, \pi]\}$$

onto itself. It leaves the boundary $\delta A = \{t = 0\} \cup \{t = \pi\}$ of $A$ invariant. If $\phi$ is
written as

$$\phi(s, t) = (s_1, t_1) = (f(s, t), g(s, t)),$$

then $\frac{\partial}{\partial t} f > 0$.

Maps of this kind are called **monotone twist maps**. We construct now a new line
through $P$ by reflecting the line segment $P_1 P$ at the normal to the curve in $P$.
This new line intersects $\Gamma$ in a new point $P_2$. Iterating this, we end up with a
sequence of points $P_n$, where $\phi(P_n) = P_{n+1}$. The set $\{P_n \mid n \in \mathbb{N}\}$ is called an
**orbit** of $P$.

An orbit is called **closed** or **periodic** if there exists $n > 0$ with $P_{i+n} = P_i$. We can
define $f$ also on the strip $\tilde{A}$ which is the covering surface

$$\tilde{A} = \mathbb{R} \times [0, \pi]$$

of $A$. For the lifted map $\tilde{\phi}$ define $\tilde{\phi}(s, 0) = 0$, $\tilde{\phi}(s, \pi) = 1$. One calls a point $P$
**periodic of type** $p/q$ with $p \in \mathbb{Z}$, $q \in \mathbb{N} \setminus \{0\}$, if $s_q = s + p, t_q = t$. In this case,

$$\lim_{n \to \infty} \frac{s_n}{n} = \frac{p}{q}$$

holds. An orbit is called **of type** $\alpha$, if

$$\lim_{n \to \infty} \frac{s_n}{n} = \alpha.$$

A first question is whether orbits of prescribed type $\alpha \in (0, 1)$ exist. We will deal
with billiards in the last chapter and outline there the connection with the calculus
of variations.

## 1.3   The accessory variational problem

In this section we learn additional necessary conditions for minimals.

**Definition.** If $\gamma^*$ is an extremal solution in $\Lambda$ and $\gamma_\epsilon = \gamma^* + \epsilon\phi$ with $\phi \in \mathrm{Lip}_0[t_1, t_2]$, we define the **second variation** as

$$
II(\phi) \;=\; \left(\frac{d^2}{(d\epsilon)^2}\right) I(\gamma_\epsilon)|_{\epsilon=0}
$$

$$
=\; \int_{t_1}^{t_2} (A\dot{\phi}, \dot{\phi}) + 2(B\dot{\phi}, \phi) + (C\phi, \phi) \, dt \, ,
$$

where $A = F_{pp}(t, x^*, \dot{x}^*), B = F_{px}(t, x^*, \dot{x}^*)$ and $C = F_{xx}(t, x^*, \dot{x}^*)$. More generally we define the symmetric bilinear form

$$
II(\phi, \psi) = \int_{t_1}^{t_2} (A\dot{\phi}, \dot{\psi}) + (B\dot{\phi}, \psi) + (B\dot{\psi}, \phi) + (C\phi, \psi) \, dt
$$

and put $II(\phi) = II(\phi, \phi)$.

It is clear that $II(\phi) \geq 0$ is a necessary condition for a minimum.

**Remark.** The symmetric bilinear form II plays the role of the **Hessian matrix** for an extremal problem on $\mathbb{R}^m$.

For fixed $\phi$, we can look at the functional $II(\phi, \psi)$ as a variational problem. It is called the **accessory variational problem**. With

$$
F(t, \phi, \dot{\phi}) = (A\dot{\phi}, \dot{\phi}) + 2(B\dot{\phi}, \phi) + (C\phi, \phi) \, ,
$$

the Euler equations to this problem are

$$
\frac{d}{dt}\left(F_{\dot{\psi}}\right) = F_\psi
$$

which are

$$
\frac{d}{dt}(A\dot{\phi} + B^T\phi) = B\dot{\phi} + C\phi \, . \tag{1.10}
$$

These equations are called the **Jacobi equations** for $\phi$.

**Definition.** Given an extremal solution $\gamma^* : t \mapsto x^*(t)$ in $\Lambda$. A point $(s, x^*(s)) \in \Omega$ with $s > t_1$ is called a **conjugate point** to $(t_1, x^*(t_1))$, if a nonzero solution $\phi \in \mathrm{Lip}[t_1, t_2]$ of the Jacobi equations (1.10) exists, which satisfies $\phi(t_1) = 0$ and $\phi(s) = 0$.
We also say, $\gamma^*$ has **no conjugate points**, if no conjugate point of $(t_1, x^*(t_1))$ exists on the open segment $\{(t, x^*(t)) \mid t_1 < t < t_2\} \subset \Omega$.

---

**Theorem 1.3.1.** *If $\gamma^*$ is a minimal, then $\gamma^*$ has no conjugate point.*

---

*Proof.* It is enough to show that $II(\phi) \geq 0$ for all $\phi \in \mathrm{Lip}_0[t_1, t_2]$ implies that no conjugate point of $(t_1, x(t_1))$ exists on the open segment $\{(t, x^*(t)) \mid t_1 < t < t_2 \}$.

Let $\psi \in \mathrm{Lip}_0[t_1, t_2]$ be a solution of the Jacobi equations, with $\psi(s) = 0$ for $s \in (t_1, t_2)$ and $\phi(\psi, \dot{\psi}) = (A\dot{\psi} + B^T \psi)\dot{\psi} + (B\dot{\psi} + C\psi)\psi$. Using the Jacobi equations, we get

$$
\begin{aligned}
\int_{t_1}^{s} \phi(\psi, \dot{\psi})\, dt &= \int_{t_1}^{s} (A\dot{\psi} + B^T \psi)\dot{\psi} + (B\dot{\psi} + C\psi)\psi \ dt \\
&= \int_{t_1}^{s} (A\dot{\psi} + B^T \psi)\dot{\psi} + \frac{d}{dt}(A\dot{\psi} + B^T \psi)\psi \ dt \\
&= \int_{t_1}^{s} \frac{d}{dt}[(A\dot{\psi} + B^T \psi)\psi]\ dt \\
&= [(A\dot{\psi} + B^T \psi)\psi]|_{t_1}^{s} = 0 \ .
\end{aligned}
$$

Because $\dot{\psi}(s) \neq 0$, the assumption $\dot{\psi}(s) = 0$ would with $\psi(s) = 0$ and the uniqueness theorem for ordinary differential equations imply that $\psi(s) \equiv 0$. This is excluded by assumption.

The Lipschitz function

$$
\tilde{\psi}(t) := \begin{cases} \psi(t), & t \in [t_1, s) \ , \\ 0, & t \in [s, t_2] \ , \end{cases}
$$

satisfies, by the above calculation, $II(\tilde{\psi}) = 0$. It is therefore also a solution of the Jacobi equation. Because we have assumed $II(\phi) \geq 0, \ \forall \phi \in \mathrm{Lip}_0[t_1, t_2]$, $\psi$ must be minimal. $\psi$ is however not $C^2$, because $\dot{\psi}(s) \neq 0$, but $\dot{\psi}(t) = 0$ for $t \in (s, t_2]$. This is a contradiction to Theorem 1.1.2. $\qquad \square$

The question now arises whether the existence of conjugate points of $\gamma$ in $(t_1, t_2)$ implies that $II(f) \geq 0$ for all $\phi \in \mathrm{Lip}_0[t_1, t_2]$. The answer is yes in the case $n = 1$. In the following, we also will deal with the one-dimensional case $n = 1$ and assume that $A, B, C \in C^1[t_1, t_2]$ and $A > 0$.

---

**Theorem 1.3.2.** *Given $n = 1, A > 0$ and an extremal solution $\gamma^* \in \Lambda$. There are no conjugate points of $\gamma$ if and only if*

$$
II(\phi) = \int_{t_1}^{t_2} A\dot{\phi}^2 + 2B\phi\dot{\phi} + C\phi^2 \ dt \geq 0, \ \forall \phi \in \mathrm{Lip}_0[t_1, t_2] \ .
$$

---

The assumption $II(\phi) \geq 0, \forall \phi \in \mathrm{Lip}_0[t_1, t_2]$ is called the **Jacobi condition**. Theorem 1.3.1 and Theorem 1.3.2 together say that a minimal satisfies the Jacobi condition in the case $n = 1$.

*Proof.* One direction has been dealt with already in the proof of Theorem 1.3.1. We still have to show how the existence theory of conjugate points for an extremal solution $\gamma^*$ leads to

$$\int_{t_1}^{t_2} A\dot{\phi}^2 + 2B\phi\dot{\phi} + C\phi^2 \, dt \geq 0, \forall \phi \in \mathrm{Lip}_0[t_1, t_2] \ .$$

First we prove this under the stronger assumption that there exists no conjugate point in $(t_1, t_2]$. We claim that there is a solution $\tilde{\phi} \in \mathrm{Lip}[t_1, t_2]$ of the Jacobi equations which satisfies $\tilde{\phi}(t) > 0, \forall t \in [t_1, t_2]$ and $\tilde{\phi}(t_1 - \epsilon) = 0$ as well as $\dot{\tilde{\phi}}(t_1 - \epsilon) = 1$ for a certain $\epsilon > 0$. One can see this as follows:

Consider a solution $\psi$ of the Jacobi equations with $\psi(t_1) = 0, \dot{\psi}(t_1) = 1$, so that by assumption the next larger root $s_2$ satisfies $s_2 > t_2$. By continuity there is $\epsilon > 0$ and a solution $\tilde{\phi}$ with $\tilde{\phi}(t_1 - \epsilon) = 0$ and $\dot{\tilde{\psi}}(t_1 - \epsilon) = 1$ and $\tilde{\phi}(t) > 0, \forall t \in [t_1, t_2]$. For such a $\tilde{\phi}$ we can apply the following **Lemma of Legendre**:

---

**Lemma 1.3.3.** *If $\psi$ is a solution of the Jacobi equations satisfying $\psi(t) > 0, \forall t \in [t_1, t_2]$, then for every $\phi \in \mathrm{Lip}_0[t_1, t_2]$ with $\xi := \phi/\psi$ we have*

$$II(\phi) = \int_{t_1}^{t_2} A\dot{\phi}^2 + 2B\phi\dot{\phi} + C\phi^2 \, dt = \int_{t_1}^{t_2} A\psi^2 \dot{\xi}^2 \, dt \geq 0 \ .$$

---

*Proof.* The following calculation goes back to Legendre. Taking the derivative of $\phi = \xi\psi$ gives $\dot{\phi} = \dot{\psi}\xi + \psi\dot{\xi}$ and therefore

$$
\begin{aligned}
II(\phi) &= \int_{t_1}^{t_2} A\dot{\phi}^2 + 2B\phi\dot{\phi} + C\phi^2 \, dt \\
&= \int_{t_1}^{t_2} (A\dot{\psi}^2 + 2B\dot{\psi}\psi + C\psi^2)\xi^2 \, dt \\
&\quad + \int_{t_1}^{t_2} (2A\psi\dot{\psi} + 2B\psi^2)\xi\dot{\xi} \, dt + \int_{t_1}^{t_2} A\psi^2\dot{\xi}^2 \, dt \\
&= \int_{t_1}^{t_2} [(A\dot{\psi} + B\psi)\dot{\psi} + \frac{d}{dt}(A\dot{\psi} + B\psi)\psi]\xi^2 \\
&\quad + (A\dot{\psi} + B\psi)\psi\frac{d}{dt}\xi^2 + A\psi^2\dot{\xi}^2 \, dt \\
&= \int_{t_1}^{t_2} \frac{d}{dt}\left((A\dot{\phi} + B\psi)\psi\xi^2\right) \, dt + \int_{t_1}^{t_2} A\psi^2\dot{\xi}^2 \, dt \\
&= (A\dot{\phi} + B\psi)\psi\xi^2|_{t_1}^{t_2} + \int_{t_1}^{t_2} A\psi^2\dot{\xi}^2 \, dt \\
&= 0 + \int_{t_1}^{t_2} A\psi^2\dot{\xi}^2 \, dt \ ,
\end{aligned}
$$

where we have used in the third equality that $\phi$ satisfies the Jacobi equations. $\square$

For the continuation of the proof of Theorem 1.3.2 we still have to deal with the case when $(t_2, x^*(t_2))$ is a conjugate point. This is Problem 6 in the exercises. $\square$

The next theorem is true only when $n = 1, A(t, x, p) > 0, \forall (t, x, p) \in \Omega \times \mathbb{R}$.

---

**Theorem 1.3.4.** *Assume $n = 1, A > 0$. For $i = 1, 2$ let $\gamma_i$ be minimals in*

$$\Lambda_i = \{\gamma : t \mapsto x_i(t) \mid x_i \in \text{Lip}[t_1, t_2], x_i(t_1) = a_i, x_i(t_2) = b_i \}.$$

*The minimals $\gamma_1$ and $\gamma_2$ intersect for $t_1 < t < t_2$ at most once.*

---

*Proof.* Assume there are two $\gamma_i$ in $\Lambda_i$ which intersect in the interior of the interval $[t_1, t_2]$ at the places $s_1$ and $s_2$ with $s_1 \neq s_2$.

We define new paths $\underline{\gamma}$ and $\overline{\gamma}$ as follows:

$$\underline{\gamma}(t) = \begin{cases} \gamma_2(t) & \text{if } t \in [t_1, s_1] \cup [s_2, t_2], \\ \gamma_1(t) & \text{if } t \in [s_1, s_2], \end{cases}$$

$$\overline{\gamma}(t) = \begin{cases} \gamma_1(t) & \text{if } t \in [t_1, s_1] \cup [s_2, t_2], \\ \gamma_2(t) & \text{if } t \in [s_1, s_2]. \end{cases}$$

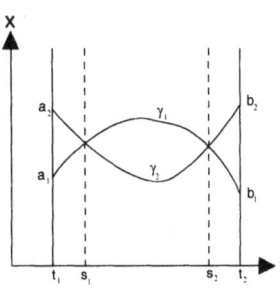

We denote also by $\tilde{\gamma}_i$ the restriction of $\gamma_i$ to $[s_1, s_2]$.
Let

$$\Lambda_0 = \{\dot{\gamma} : t \mapsto x(t), x(t) \in \text{Lip}[s_1, s_2], \ x(s_i) = x_1(s_i) = x_2(s_i) \}.$$

In this class we have $I(\tilde{\gamma}_1) = I(\tilde{\gamma}_2)$ because both $\gamma_1$ and $\gamma_2$ are minimal. This means

$$I(\overline{\gamma}) = I(\gamma_1) \text{ in } \Lambda_1,$$
$$I(\underline{\gamma}) = I(\gamma_2) \text{ in } \Lambda_2.$$

Therefore $\tilde{\gamma}$ is minimal in $\Lambda_1$ and $\gamma_2$ is minimal in $\Lambda_2$. This contradicts the regularity theorem. The curves $\overline{\gamma}$ and $\underline{\gamma}$ can therefore not be $C^2$ because $\gamma_1$ and $\gamma_2$ intersect transversally as a consequence of the uniqueness theorem for ordinary differential equations. $\square$

**Application: The Sturm theorems.**

---

**Corollary 1.3.5.** *If $s_1$ and $s_2$ are two successive roots of a solution $\phi \neq 0$ of the Jacobi equation, then every solution which is linearly independent of $\phi$ has exactly one root in the interval $(s_1, s_2)$.*

---

**Corollary 1.3.6.** *If $q(t) \leq Q(t)$ and*

$$\ddot{\phi} + q\phi = 0 \, ,$$
$$\ddot{\Phi} + Q\Phi = 0 \, ,$$

*and $s_1, s_2$ are two successive roots of $\Phi$, then $\phi$ has at most one root in $(s_1, s_2)$.*

The proof the of Sturm theorems is an exercise (see Exercise 7).

## 1.4   Extremal fields for n=1

In this section we derive sufficient conditions for minimality in the case $n = 1$. We will see that the Euler equations, the assumption $F_{pp} > 0$ and the Jacobi conditions are sufficient for a local minimum. Since all these assumptions are of local nature, one can not expect more than one local minimum. If we talk about a local minimum, this is understood with respect to the topology on $\Lambda$. In the $C^0$ topology on $\Lambda$, the distance of two elements $\gamma_1 : t \mapsto x_1(t)$ and $\gamma_2 : t \mapsto x_2(t)$ is given by

$$d(\gamma_1, \gamma_2) = \max_{t \in [t_1, t_2]} \{ |x_1(t) - x_2(t)| \} \, .$$

A neighborhood of $\gamma^*$ in this topology is called a **wide neighborhood** of $\gamma$. A different possible topology on $\Lambda$ would be the $C^1$ topology, in which the distance of $\gamma_1$ and $\gamma_2$ is measured by

$$d_1(\gamma_1, \gamma_2) = \sup_{t \in [t_1, t_2]} \{ |x_1(t) - x_2(t)| + |\dot{x}_1(t) - \dot{x}(t)| \} \, .$$

An open set containing $\gamma^*$ is then called a **narrow neighborhood** of $\gamma^*$.

**Definition.**   $\gamma^* \in \Lambda$ is called a **strong minimum** in $\Lambda$, if $I(\gamma) \geq I(\gamma^*)$ for all $\gamma$ in a wide neighborhood of $\gamma*$.
$\gamma^* \in \Lambda$ is called a **weak minimum** in $\Lambda$, if $I(\gamma) \geq I(\gamma^*)$ for all $\gamma$ in a narrow neighborhood of $\gamma^*$.

We will see that under the assumption of the Jacobi condition, a field of extremal solutions can be found which cover a wide neighborhood of the extremal solutions $\gamma^*$.

**Definition.**   An **extremal field** in $\Omega$ is a vector field $\dot{x} = \psi(t, x), \psi \in C^1(\Omega)$ which is defined in a wide neighborhood $\mathcal{U}$ of an extremal solution and which has the property that every solution $x(t)$ of the differential equation $\dot{x} = \psi(t, x)$ is also a solution of the Euler equations.

**Examples**.

1) $F = \frac{1}{2}p^2$ has the Euler equation $\ddot{x} = 0$ and the extremal field: $\dot{x} = \psi(t,x) = c = const.$

2) $F = \sqrt{1 + p^2}$ has the Euler equations $\ddot{x} = 0$ with a solution $x = \lambda t$. The equation $\dot{x} = \psi(t,x) = x/t$ defines an extremal field for $t > 0$.

3) For the geodesics on a torus embedded in $\mathbb{R}^3$, the Clairaut angle $\phi$ satisfies the equation $r \sin(\phi) = c$ with $-(a-b) < c < (a-b)$. This angle defines an extremal field. (See Exercise 12).

---

**Theorem 1.4.1.** $\psi = \psi(t,x)$ *defines an extremal field in $\mathcal{U}$ if and only if for all $\gamma \in \mathcal{U}$ and $\gamma : t \mapsto x(t)$ one has*

$$D_\psi F_p = F_x$$

*for $p = \psi(t,x)$, where $D_\psi := \partial_t + \psi\partial_x + (\psi_t + \psi\psi_x)\partial_p$.*

---

*Proof.* $\psi$ defines an extremal field if and only if for all $\gamma \in \mathcal{U}$, $\gamma : t \mapsto x(t)$

$$\frac{d}{dt}F_p(t,x,p) = F_x(t,x,p)$$

for $p = \dot{x} = \psi(t,x(t))$. We have

$$(\partial_t + \dot{x}\partial_x + \frac{d}{dt}\psi(t,x(t))\partial_p)F_p = F_x ,$$
$$(\partial_t + \psi\partial_x + (\psi_t + \psi_x\psi)\partial_p)F_p = F_x . \qquad \square$$

---

**Theorem 1.4.2.** *If $\gamma^*$ can be embedded in an extremal field in a wide neighborhood $\mathcal{U}$ of $\gamma^*$ and $F_{pp}(t,x,p) \geq 0$ for all $(t,x) \in \Omega$ and for all $p$, then $\gamma^*$ is a strong minimal. If $F_{pp}(t,x,p) > 0$ for all $(t,x) \in \Omega$ and for all $p$, then $\gamma^*$ is a unique strong minimal.*

---

*Proof.* Let $\mathcal{U}$ be a wide neighborhood of $\gamma^*$ and let $F_{pp}(t,x,p) \geq 0$ for $(t,x) \in \Omega, \forall p$. We show that $I(\gamma^*) \geq I(\gamma)$ for all $\gamma \in \mathcal{U}$. Let for $\gamma \in C^2(\Omega)$

$$\tilde{F}(t,x,p) = F(t,x,p) - g_t - g_x p ,$$
$$\tilde{I}(\gamma) = \int_{t_1}^{t_2} \tilde{F}(t,x,p) \, dt = I(\gamma) - g(t,x)\Big|_{(t_1,b)}^{(t_2,a)} .$$

We look for $\gamma \in C^2$ so that

$$\tilde{F}(t, x, \psi(t, x)) = 0 \ ,$$
$$\tilde{F}(t, x, p) \geq 0, \ \forall p \ .$$

(This means that every extremal solution of the extremal field is a minimal one!)

Such an $\tilde{F}$ defines a variational problem which is equivalent to the one defined by $F$ because $\tilde{F}_p = 0$ for $p = \psi(t, x)$.

We consider now the two equations

$$
\begin{aligned}
g_x &= F_p(t, x, \psi) \ , \\
g_t &= F(t, x, \psi) - F_p(t, x, \psi)\psi \ ,
\end{aligned}
$$

which are called the **fundamental equations of the calculus of variations**. They form a system of partial differential equations of the form

$$
\begin{aligned}
g_x &= a(t, x) \ , \\
g_t &= b(t, x) \ .
\end{aligned}
$$

These equations have solutions if $\Omega$ is simply connected and if the integrability condition $a_t = b_x$ is satisfied (if the curl of a vector field in a simply connected region vanishes, then the vector field is a gradient field). Then $g$ can be computed as a (path independent) line integral

$$g = \int a(t, x) \ dx + b(t, x) \ dt \ . \qquad \qquad \square$$

We now interrupt the proof for a lemma.

---

**Lemma 1.4.3.**  *The compatibility condition $a_t = b_x$:*

$$\frac{\partial}{\partial t} F_p(t, x, \psi(t, x)) = \frac{\partial}{\partial x}(F - \psi F_p)(t, x, \psi(t, x))$$

*is true if and only if $\psi$ is an extremal field.*

---

*Proof.* This is a calculation. One has to consider that

$$a(t, x) = F_p(t, x, \psi(t, x))$$

and that

$$b(t, x) = (F - \psi F_p)(t, x, \psi(t, x))$$

are functions of the two variables $t$ and $x$, while $F$ is a function of three variables $t, x, p$, where $p = \psi(t, x)$. We write $\partial_t F, \partial_x F$ and $\partial_p F$, for the derivatives of $F$ with respect to the first, the second and the third variables. We write $\frac{\partial}{\partial t}F(t, x, \psi(t, x))$ rsp. $\frac{\partial}{\partial x}F(t, x, \psi(t, x))$, if $p = \psi(t, x)$ is a function of the independent random variables $t$ and $x$. Therefore

$$\frac{\partial}{\partial t}a(t, x) \ = \ \frac{\partial}{\partial t}F_p(t, x, \psi(t, x)) = F_{pt} + \psi_t F_{pp} \qquad (1.11)$$

$$= \ (\partial_t + \psi_t \partial_p)F_p \ , \qquad (1.12)$$

and because

$$\frac{\partial}{\partial x}F_p(t, x, \psi(t, x)) = F_{px} + \psi_x F_{pp} = (\partial_x + \psi_x \partial_p)F_p \ ,$$

also

$$\frac{\partial}{\partial x}b(t, x) \ = \ \frac{\partial}{\partial x}[F(t, x, \psi(t, x)) - \psi(t, x)F_p(t, x, \psi(t, x))] \qquad (1.13)$$

$$= \ (\partial_x + \psi_x \partial_p)F - (\psi_x F_p + \psi F_{px} + \psi \psi_x F_{pp}) \qquad (1.14)$$

$$= \ F_x - (\psi_x + \psi \partial_x + \psi \psi_x \partial_p)F_p \ . \qquad (1.15)$$

(1.11) and (1.13) together give

$$\frac{\partial}{\partial x}b - \frac{\partial}{\partial t}a \ = \ F_x - (\partial_t + \psi \partial_x + (\psi_t + \psi \psi_x)\partial_p)F_p$$

$$= \ F_x - D_\psi F \ .$$

According to Theorem 1.4.1, the relation $\partial_x b - \partial_t a = 0$ holds if and only if $\psi$ defines an extremal field. $\qquad\square$

Continuation of the proof of Theorem 1.4.2:

*Proof.* With this lemma, we have found a function $g$ which itself can be written as a path-independent integral

$$g(t, x) = \int_{(t_1, a)}^{(t, x)} (F - \psi F_p) \, dt' + F_p \, dx'$$

called a **Hilbert invariant integral**. For every curve $\gamma : t \mapsto x(t)$ one has:

$$I(\gamma) = \int_\gamma F \, dt = \int_\gamma F \, dt - F_p \dot{x} \, dt + F_p \, dx \ . \qquad (1.16)$$

Especially for the path $\gamma^*$ of the extremal field $\dot{x} = \psi(t, x)$, one has

$$I(\gamma^*) = \int_{\gamma^*} (F - \psi F_p) \, dt + F_p \, dx \ . \qquad (1.17)$$

For every $\gamma \in \Lambda$ the difference of (1.17) with (1.16) gives

$$\begin{aligned} I(\gamma) - I(\gamma^*) &= \int_\gamma F(t,x,\dot{x}) - F(t,x,\psi) - (\dot{x} - \psi)F_p(t,x,\psi)\, dt \\ &= \int_\gamma E(t,x,\dot{x},\psi)\, dt\,, \end{aligned}$$

where $E(t,x,p,q) = F(t,x,p) - F(t,x,q) - (p-q)F_p(t,x,q)$ is called the **Weierstrass excess function** or shortly the **Weierstrass E-function**. By the intermediate value theorem there is a value $\bar{q} \in [p,q]$ with

$$E(t,x,p,q) = \frac{(p-q)^2}{2} F_{pp}(t,x,\bar{q}) \geq 0\,.$$

This inequality is strict if $F_{pp} > 0$ and $p \neq q$. Therefore $I(\gamma) - I(\gamma^*) \geq 0$ and if $F_{pp} > 0$, then $I(\gamma) > I(\gamma^*)$ for $\gamma \neq \gamma^*$. In other words $\gamma^*$ is a unique strong minimal. $\qquad\qquad\qquad\square$

The Euler equations, the Jacobi condition and the condition $F_{pp} \geq 0$ are sufficient for a strong local minimum:

---

**Theorem 1.4.4.** *Let $\gamma^*$ be an extremal with no conjugate points. Assume $F_{pp} \geq 0$ on $\Omega$ and let $\gamma^*$ be embedded in an extremal field. It is therefore a strong minimal. If $F_{pp} > 0$ on $\Omega$, then $\gamma^*$ is a unique minimal.*

---

*Proof.* We construct an extremal field which contains $\gamma^*$ and make Theorem 1.4.2 applicable.

Choose $\tau < t_1$ close enough to $t_1$, so that all solutions $\phi$ of the Jacobi equations with $\phi(\tau) = 0$ and $\dot{\phi}(\tau) \neq 0$ are nonzero on $(\tau, t_2]$. This is possible by continuity. We construct now a field $x = u(t,\eta)$ of solutions to the Euler equations, so that for small enough $|\eta|$,

$$\begin{aligned} u(\tau,\eta) &= x^*(\tau)\,, \\ \dot{u}(\tau,\eta) &= \dot{x}^*(\tau) + \eta\,. \end{aligned}$$

This can be achieved by the existence theorem for ordinary differential equations. We show that for some $\delta > 0$ with $|\eta| < \delta$, these extremal solutions cover a wide neighborhood of $\gamma^*$. To do so we prove that $u_\eta(t,0) > 0$ for $t \in (\tau, t_2]$.

If we differentiate the Euler equations

$$\frac{d}{dt}F_p(t,u,\dot{u}) = F_x(t,u,\dot{u})$$

at $\eta = 0$ with respect to $\eta$ we get

$$\frac{d}{dt}(A\ddot{u}_\eta + B\dot{u}_\eta) = B\dot{u}_\eta + Cu_\eta$$

and see that $\phi = u_\eta$ is a solution of the Jacobi equations. With the claim $u_\eta(t,0) > 0$ for $t \in [t_1, t_2]$ we obtained the statement at the beginning of the proof.

From $u_\eta(t,0) > 0$ in $(\tau, t_2]$ follows, with the implicit function theorem, that for $\eta$ in a neighborhood of zero, there is an inverse function $\eta = v(t,x)$ of $x = u(t,\eta)$ which is $C^1$ and for which the equation

$$0 = v(t, x^*(t))$$

holds. Especially, the $C^1$ function ($u_t$ and $v$ are $C^1$)

$$\psi(t,x) = u_t(t, v(t,x))$$

defines an extremal field $\psi$,

$$\dot{x} = \psi(t,x)$$

which is defined in a neighborhood of $\{(t, x^*(t)) \mid t_1 \le t \le t_2\}$. Of course every solution of $\dot{x} = \psi(t,x)$ in this neighborhood is given by $x = u(t,h)$ so that every solution of $\dot{x} = \psi(t,x)$ is an extremal. $\qquad \square$

## 1.5 The Hamiltonian formulation

The Euler equations

$$\frac{d}{dt}F_{p_j} = F_{x_j} ,$$

which an extremal solution $\gamma$ in $\Lambda$ has to satisfy, form a system of second order differential equations. If $\sum_{i,j} F_{p_i p_j}\xi^i\xi^j > 0$ for $\xi \ne 0$, the **Legendre transformation**

$$l : \Omega \times R^n \to \Omega \times \mathbb{R}^n, (t,x,p) \mapsto (t,x,y)$$

is defined, where $y_j = F_{p_j}(t,x,p)$ is uniquely invertible. It is in general not surjective. A typical example of a not surjective case is

$$F = \sqrt{1+p^2}, \quad y = \frac{p}{\sqrt{1+p^2}} \in (-1,1) .$$

The Legendre transformation relates the Lagrange function with the **Hamilton function**

$$H(t,x,y) = (y,p) - F(t,x,p) ,$$

where

$$p = H_y(t,x,y) .$$

We have $H_{yy}(t, x, y) = p_y = y_p^{-1} = F_{pp}^{-1} > 0$ and the Euler equations become, after a Legendre transformation, the **Hamilton differential equations**

$$\dot{x}_j = H_{y_j} ,$$
$$\dot{y}_j = -H_{x_j} .$$

They form a system of first order differential equations. One can write these Hamilton equations again as Euler equations using the **action integral**

$$S = \int_{t_1}^{t_2} y\dot{x} - H(t, x, y) \, dt .$$

This was **Cartan's approach** to this theory. The differential form

$$\alpha = ydx - Hdt = dS$$

is called the **integral invariant of Poincaré–Cartan**. The above action integral is of course the Hilbert invariant Integral which we met in the third section.

If the Legendre transformation is surjective we call $\Omega \times \mathbb{R}^n$ **the phase space**. It is important that $y$ is now independent of $x$ so that the differential form $\alpha$ does not depend only on the $(t, x)$ variables: it is also defined in the phase space $\Omega \times \mathbb{R}^n$.

If $n = 1$, the phase space is three dimensional. For a function $h : (t, x) \mapsto h(t, x)$ the graph

$$\Sigma = \{(t, x, y) \in \Omega \times \mathbb{R}^n \mid y = h(t, x) \}$$

is a two-dimensional surface.

**Definition.** The surface $\Sigma$ is called **invariant** under the flow of H, if the vector field

$$X_H = \partial_t + H_y\partial_x - H_x\partial_y$$

is tangent to $\Sigma$.

---

**Theorem 1.5.1.** *Let* $(n = 1)$. *If* $\dot{x} = \psi(t, x)$ *is an extremal field for* $F$, *then*

$$\Sigma = \{(t, x, y) \in \Omega \times \mathbb{R} \mid y = F_p(t, x, \psi(t, x)) \}$$

*is* $C^1$ *and invariant under the flow of H. On the other hand, if* $\Sigma$ *is a surface which is invariant under the flow of H and has the form*

$$\Sigma = \{(t, x, y) \in \Omega \times \mathbb{R} \mid y = h(t, x) \} ,$$

*where* $h \in C^1(\Omega)$, *then the vector field* $\dot{x} = \psi(t, x)$ *defined by*

$$\psi = H_y(t, x, h(t, x))$$

*is an extremal field.*

---

*Proof.* We assume first that an extremal field $x = \psi(t, x)$ for $F$ is given. Then according to Theorem 1.11,

$$D_\psi F_p = F_x$$

and by the lemma in the proof of Theorem 1.4.2 this is the case if and only if there exists a function $g$ which satisfies the **fundamental equations**

$$
\begin{aligned}
g_x(t, x) &= F_p(t, x, \psi), \\
g_t(t, x) &= F(t, x, \psi) - \psi F_p(t, x, \psi) = -H(t, x, g_x).
\end{aligned}
$$

The surface

$$\Sigma = \{ (t, x, p) \mid y = g_x(t, x, \psi) \}$$

is invariant under the flow of $H$:

$$
\begin{aligned}
X_H(y - g_x) &= [\partial_t + H_y \partial_x - H_x \partial_y](y - g_x) \\
&= -H_y g_{xx} - g_{xt} - H_x \\
&= -\partial_x[g_t + H(t, x, g_x)] = 0.
\end{aligned}
$$

On the other hand, if

$$\Sigma = \{ (t, x, p) \mid y = h(t, x) \}$$

is invariant under the flow of $H$, then by definition

$$
\begin{aligned}
0 = X_H(y - h(t, x)) &= [\partial_t + H_y \partial_x - H_x \partial_y](y - h(t, x)) \\
&= -H_y h_x - h_t - H_x \\
&= -\partial_x[g_t + H(t, x, h)]
\end{aligned}
$$

with a function $g(t, x) = \int_a^x h(t, x') \, dx'$ satisfying the **Hamilton–Jacobi equations**

$$
\begin{aligned}
g_x &= h(t, x) = y = F_p(t, x, \dot{x}), \\
g_t &= -H(t, x, g_x).
\end{aligned}
$$

This means that $\dot{x} = g_x(t, x) = H_y(t, x, h(x, y))$ defines an extremal field. $\square$

Theorem 1.5.1 tells us that instead of considering extremal fields we can look at surfaces which are given as the graph of $g_x$, where $g$ is a solution of the **Hamilton–Jacobi equation**

$$g_t = -H(t, x, g_x).$$

They can be generalized to $n \geq 1$: We look for $g \in C^2(\Omega)$ at the manifold $\Sigma := \{ (t, x, y) \in \Omega \times \mathbb{R}^n \mid y_j = g_{x_j} \}$, where

$$g_t + H(t, x, g_x) = 0.$$

The following result holds:

**Theorem 1.5.2.** *a)* $\Sigma$ *is invariant under* $X_H$.
*b) The vector field* $\dot{x} = \psi(t, x)$, *with* $\psi(t, x) = H_y(t, x, g_x)$ *defines an extremal field for* $F$.
*c) The Hilbert integral* $\int F + (\dot{x} - \psi)F \, dt$ *is path independent.*

The verification of these theorems is done as before in Theorem 1.5.1. One has to consider that in the case $n > 1$ **not every** field $\dot{x} = \psi(t, x)$ of extremal solutions can be represented in the form $\psi = H_y$. The necessary assumption is the solvability of the fundamental equations

$$
\begin{aligned}
g_t &= F(t, x, \psi) - \sum_{j=1}^{n} \psi_j F_{p_j}(t, x, \psi) \, , \\
g_x &= F_{p_j}(t, x, \psi) \, .
\end{aligned}
\tag{1.18}
$$

From the $n(n+1)/2$ compatibility conditions which have to be satisfied, only the $n(n-1)/2$ assumptions

$$
\partial_{x_k} F_{p_j}(t, x, \psi) = \partial_{x_j} F_{p_k}(t, x, \psi)
\tag{1.19}
$$

are necessary. Additionally, the $n$ conditions

$$
D_\psi F_{p_j}(t, x, \psi) = F_{x_j}(t, x, \psi)
$$

hold. They express that solutions of $\dot{x} = \psi$ are extremal.

**Definition.** A vector field $\dot{x} = \psi(t, x)$ is called a **Mayer field** if there is a function $g(t, x)$ which satisfies the fundamental equations (1.18).

We have seen that a vector field is a Mayer field if and only if it is an extremal field which satisfies the **compatibility conditions** (1.19). Equivalently, the differential form

$$
\alpha = \sum_j y_j dx_j - H(t, x, y) \, dt
$$

is closed on $\Sigma = \{(t, x, y) \mid y = h(t, x)\}$:

$$
d\alpha|_\Sigma = d\left[ \sum_j h_j dx_j - H(t, x, h) dt \right] = 0 \, .
$$

Because $\Omega$ is simply connected this is equivalent to exactness $\alpha|_\Sigma = dg$ or

$$
\begin{aligned}
h_j &= g_{x_j} \, , \\
-H(t, x, h) &= g_t \, ,
\end{aligned}
$$

which is, with the Legendre transformation, equivalent to the fundamental equations

$$F_p(t, x, \psi) = g_x ,$$
$$F(t, x, \psi) - \psi F_p = g_t .$$

In this way, a Mayer field defines a manifold as the graph of a function $y = h(t, x)$ in such a way that $d\alpha = 0$ on $g = h$.

In invariant terminology an $n$-dimensional submanifold of a $(2n + 1)$-dimensional manifold with a 1-form $\alpha$ is called a **Legendre manifold**, if $d\alpha$ vanishes there. (See [3] Appendix 4K).

**Geometric interpretation of $g$.**
A Mayer field given by a function $g = g(t, x)$ which satisfies $g_t + H(t, x, g_x) = 0$ is

$$\dot{x} = H_y(t, x, g_x) = \psi(t, x) .$$

This has the following geometric significance:

The manifolds $g \equiv const$, as for example the manifolds $g \equiv A$ and $g \equiv B$, are equidistant with respect to $\int F\, dt$ in the sense that along an extremal solution $\gamma : t \mapsto x(t)$ with $x(t_A) \in \{g = A\}$ and $x(t_B) \in \{g = B\}$ one has

$$\int_{t_A}^{t_B} F(t, x(t), \psi(t, x(t)))\, dt = B - A .$$

Therefore

$$\frac{d}{dt} g(t, x(t)) = g_t + \psi g_x = F - \psi F_p(t, x, \psi) + \psi F_p(t, x, \psi) = F(t, x, \psi) ,$$

and

$$\int_{t_A}^{t_B} F(t, x(t), \psi(t, x(t)))\, dt = \int_{t_A}^{t_B} \frac{d}{dt} g(t, x(t))\, dt = g(t, x(t))|_{t_A}^{t_B} = B - A .$$

Because these are minimals, $\int F(t, x, \psi(t, x)\, dt$ measures a distance between the manifolds $g = const$. The latter are also called **wave fronts**, an expression which has its origin in optics, where $F(x, p) = \eta(x)\sqrt{1 + |p|^2}$ and $\eta(x)$ is called the **refraction index**. The function $g$ is often denoted by $S = S(t, x)$. The Hamilton–Jacobi equation

$$S_t + H(x, S_x) = 0$$

has in this case the form

$$S_t^2 + |S_x|^2 = \eta^2 .$$

Therefore

$$F_p = \eta \frac{p}{\sqrt{1+|p|^2}} = y, p = \frac{y}{\sqrt{\eta^2 - |y|^2}},$$

$$H = pF_p - F = -\eta/\sqrt{\eta^2 - |y|^2} = -\sqrt{\eta^2 - |y|^2}$$

and consequently $S_t + H(x, S_x) = S_t - \sqrt{\eta^2 - S_x^2} = 0$ holds. The corresponding extremal field

$$\dot{x} = \psi(t, x) = H_y(t, S_x) = \frac{-S_x}{\sqrt{\eta^2 - |S_x|^2}} = \frac{-S_x}{S_t}$$

is in the $(t, x)$-space orthogonal to $S(t, x) = const.$:

$$(\dot{t}, \dot{x}) = (1, \dot{x}) = \lambda(S_t, S_x)$$

with $\lambda = S_t^{-1}$. 'Light rays are orthogonal to wave fronts'.

## 1.6   Exercises to Chapter 1

1) Show that in example 4) of section 1.1, the metric $g_{ij}$ has the form given there.

2) In Euclidean three-dimensional space, a surface of revolution is given in cylindrical coordinates as

$$f(z, r) = 0 .$$

The local coordinates on the surface of revolution are $z$ and $\phi$. The surface is defined by the function $r = r(z)$ giving the distance from the axes of rotation.

a) Show that the Euclidean metric on $\mathbb{R}^3$ induces the metric on the cylinder given by

$$ds^2 = g_{11}dz^2 + g_{22}d\phi^2$$

with

$$g_{11} = 1 + (\frac{dr}{dz})^2, \ g_{22} = r^2(z) .$$

b) Let $F((\phi, z), (\dot{\phi}, \dot{z})) = \frac{1}{2}(g_{zz}\dot{z}^2 + r^2(z)\dot{\phi}^2)$. Show that along a geodesic the functions

$$p_\phi := \frac{\partial F}{\partial \dot{\phi}} r^2 \dot{\phi}, p_z := \frac{\partial F}{\partial \dot{z}} = g_{11}\dot{z}$$

are constant.

**Hint.** Proceed as in example 4) and work with $z$ and $\phi$ as 'time parameter'.

c) Denote by $e_z$ and $e_\phi$ the standard basis vectors on the cylinder $\mathbb{R} \times \mathbb{T}$ and a point on the cylinder by $(z, \phi)$. The angle $\psi$ between $e_\phi$ and the tangent vector $v = (\dot{z}, \dot{\phi})$ at the geodesic is given by

$$\cos(\psi) = (v, e_\phi) / \sqrt{(v, v)(e_\phi, e_\phi)} \, .$$

Show that $r \cos(\psi) = p_\phi / \sqrt{F}$, and that consequently the **theorem of Clairaut** holds, which says that $r \cos(\psi)$ is constant along every geodesic on the surface of revolution.

d) Show that the geodesic flow on a surface of revolution is completely integrable. Find the formulas for $\phi(t)$ and $z(t)$.

3) Show that there exists a triangle inscribed into a smooth convex billiards which has maximal length. (In particular, this triangle does not degenerate to a 2-gon.) Show that this triangle is a closed periodic orbit for the billiards.

4) Prove that the billiards in a circle has for every $p/q \in (0, 1)$ periodic orbits of type $\alpha = p/q$.

5) Let $A > 0$ and $A, B, C \in C^1[t_1, t_2]$. Consider the linear differential operator

$$L\Phi = \frac{d}{dt}(A\dot{\Phi} + B\Phi) - (B\dot{\Phi} + C\Phi) \, .$$

Prove that for $\psi > 0, \psi \in C^1[t_1, t_2]$, $\zeta \in C^1[t_1, t_2]$ the identity

$$L(\zeta\psi) = \psi^{-1} \frac{d}{dt}(A\psi^2 \dot{\zeta}) + \zeta L(\psi)$$

holds. Especially for $L\psi = 0, \psi > 0$ one has

$$L(\zeta\psi) = \psi^{-1} \frac{d}{dt}(A\psi^2 \dot{\zeta}) \, .$$

Compare this formula with the Legendre transformation for the second variation.

6) Complete the proof of Theorem 1.3.2 using the Lemma of Legendre. One has still to show that for all $\phi \in \mathrm{Lip}_0[t_1, t_2]$ the inequality

$$II(\phi) = \int_{t_1}^{t_2} A\dot{\phi}^2 + 2B\phi\dot{\phi} + C\phi^2 \, dt \geq 0$$

holds if $(t_2, x^*(t_2))$ is the nearest conjugate point to $(t_1, x^*(t_1))$. Choose for every small enough $\epsilon > 0$ a $C^1$ function $\eta_\epsilon$, for which

$$\eta_\epsilon(t) = \begin{cases} 0 & t \in (-\infty, t_1 + \epsilon/2) \cup (t_2 - \epsilon/2, \infty) \, , \\ 1 & t \in [t_1 + \epsilon, t_2 - \epsilon] \, , \end{cases}$$

$$\dot{\eta}_\epsilon(t) = O(\epsilon^{-1}), \epsilon \to 0 \, ,$$

and show then that

a) $II(\eta_\epsilon \phi) \geq 0$, $\forall \epsilon$ small enough,
b) $II(\eta_\epsilon \phi) \to II(\phi)$ for $\epsilon \to 0$.

7) Prove the Sturm theorems (Corollaries 1.3.5 and 1.3.6).

8) Let $F \in C^2(\Omega \times \mathbb{R})$ be given in such a way that every $C^2$ function $t \mapsto x(t)$, $(t, x(t)) \in \Omega$ satisfies the Euler equation

$$\frac{d}{dt} F_p(t, x, \dot{x}) = F_x(t, x, \dot{x}) \ .$$

Prove that if $\Omega$ is simply connected, $F$ must have the form

$$F(t, x, p) = g_t + g_x p$$

with $g \in C^1(\Omega)$.

9) Show that for all $x \in \text{Lip}_0[0, a]$

$$\int_0^a \dot{x}^2 - x^2 \ dt \geq 0$$

if and only if $|a| \leq \pi$.

10) Show that $x \equiv 0$ (the function which is identically 0) is not a strong minimal for

$$\int_0^1 F(t, x, \dot{x}) \ dt = \int_0^1 (\dot{x}^2 - \dot{x}^4) \ dt, x(0) = x(1) = 0 \ .$$

11) Determine the distance between the conjugate points of the geodesics $v \equiv 0$ in Example 4) and show, that on the geodesic $v \equiv 1/2$, there are no conjugate points.

**Hint.** Linearize the Euler equations for $F = \sqrt{\frac{a}{b} + \cos(2\pi v))^2 + (v')^2}$.

12) Show that the geodesic in example 4) which is given by $I = r \sin(\psi)$ defines an extremal field if $-(a - b) < c < a - b$. Discuss the geodesic for $c = a - b$, for $a - b < c < a + b$ and for $c = a + b$.

# Chapter 2

# Extremal fields and global minimals

## 2.1 Global extremal fields

The two-dimensional torus has the standard representation $\mathbb{T}^2 = \mathbb{R}^2/\mathbb{Z}^2$. We often will work on its covering surface $\mathbb{R}^2$, where everything is invariant under its fundamental group $\mathbb{Z}^2$. In this chapter we deal with the variational principle $\int F(t,x,p)\, dt$ on $\mathbb{R}^2$, where $F$ is assumed to satisfy the following properties:

i) $F \in C^2(\mathbb{T}^2 \times \mathbb{R}^2)$:

$$
\begin{aligned}
a) \quad & F \in C^2(\mathbb{R}^3)\,, \\
b) \quad & F(t+1,x,p) = F(t,x+1,p) = F(t,x,p)\,.
\end{aligned}
\tag{2.1}
$$

ii) $F$ has quadratic growth: There exist $\delta > 0$, $c > 0$ such that

$$
\begin{aligned}
c) \quad & \delta \leq F_{pp} \leq \delta^{-1}\,, \\
d) \quad & |F_x| \leq c(1 + p^2)\,, \\
e) \quad & |F_{tp}| + |F_{px}| \leq c(1 + |p|)\,.
\end{aligned}
\tag{2.2}
$$

Because of $F_t = -H_t$, $F_x = -H_x$ and $F_{pp} = H_{yy}^{-1}$ these assumptions appear in the Hamiltonian formulation as follows:

i) $H \in C^2(\mathbb{T}^2 \times \mathbb{R}^2)$:

$$
\begin{aligned}
a) \quad & H \in C^2(\mathbb{R}^3)\,, \\
b) \quad & H(t+1,x,y) = H(t,x+1,y) = H(t,x,y)\,.
\end{aligned}
\tag{2.3}
$$

ii) $H$ has quadratic growth: There exist $\delta > 0$, $c > 0$ such that

$$
\begin{aligned}
c) & \quad \delta \le H_{yy} \le \delta^{-1}, \\
d) & \quad |H_x| \le c(1 + y^2), \\
e) & \quad |H_{ty}| + |F_{yx}| \le c(1 + |y|).
\end{aligned}
\tag{2.4}
$$

**Example. Nonlinear pendulum.**

Let $V(t, x) \in C^2(\mathbb{T}^2)$ be defined as

$$
V(t, x) = (g(t)/(2\pi)) \cos(2\pi x)
$$

and $F = p^2/2 + V(t, x)$. The Euler equation

$$
\ddot{x} = g(t) \sin(2\pi x)
\tag{2.5}
$$

is a differential equation which describes a pendulum, where the gravitational acceleration $g$ is periodic and time dependent. A concrete example would be the tidal force of the moon. The linearized equation of (2.5) is called the **Hills equation**

$$
\ddot{x} = g(t)x
$$

and has been investigated in detail, especially in the case $g(t) = -\omega^2(1 + \epsilon \cos(2\pi t))$, where Hills equation is called the **Mathieu equation**. One is interested in the stability of the system in dependence on the parameters $\omega$ and $\epsilon$. One could ask for example whether the weak tidal force of the moon could pump up a pendulum on the earth, if the motion of the pendulum is without friction.

The just encountered stability question is central to the general theory.

**Definition.** A **global extremal field** on the torus is a vector field $\dot{x} = \psi(t, x)$ with $\psi \in C^1(\mathbb{T}^2)$, for which every solution $x(t)$ is extremal: $D_\psi F_p - F_x|_{p=\psi} = 0$.

Are there such extremal fields at all?

**Example.** The free nonlinear pendulum.
If the gravitational acceleration $g(t) = g$ is constant, there is an extremal field. In this case, $F$ is autonomous, and according to Theorem 1.1.5,

$$
E = pF_p - F = p^2/2 - V(x) = const.
$$

so that for $E > \max\{V(x) \mid x \in T^1\}$ an extremal field is given by

$$
\dot{x} = \psi(t, x) = \sqrt{2(E - V(x))}.
$$

The problem is thus integrable and explicit solutions can be found using an **elliptic integral**.

The existence of an extremal field is equivalent to stability. Therefore, we know with Theorem 1.5.1 that in this case, the surfaces

$$\Sigma = \{(t, x, y) \mid y = F_p(t, x, \psi(t, x))\}$$

are invariant under the flow of $X_H$.

The surface $\Sigma$ is an **invariant torus** in the phase space $\mathbb{T}^2 \times \mathbb{R}^2$. The question of the existence of invariant tori is subtle and part of the so called **KAM theory**. We will come back to it in the last chapter.

**Definition.** An extremal solution $x = x(t)$ is called a **global minimal**, if

$$\int_{\mathbb{R}} F(t, x + \phi, \dot{x} + \dot{\phi}) - F(t, x, \dot{x}) \, dt \geq 0$$

for all $\phi \in \text{Lip}_{comp}(\mathbb{R}) = \{\phi \in \text{Lip}(\mathbb{R}) \text{ with compact support.}\}$

**Definition.** A curve $\gamma : t \mapsto x(t)$ has a **self intersection** in $\mathbb{T}^2$, if there exists $(j, k) \in \mathbb{Z}^2$ such that the function $x(t + j) - k - x(t)$ changes sign.

In order that a curve has no self intersection we must have for all $(j, k) \in \mathbb{Z}^2$ either $x(t + j) - k - x(t) > 0$ or $x(t + j) - k - x(t) = 0$ or $x(t + j) - k - x(t) < 0$.

---

**Theorem 2.1.1.** *If $\psi \in C^1(\mathbb{T})$ is an extremal field, then every solution of $\dot{x} = \psi(t, x)$ is a global minimal and has no self intersections on the torus.*

---

*Proof.* Assume $\overline{\gamma} : t \mapsto \overline{x}(t)$ is a solution of the extremal field $\dot{x} = \psi(t, x)$. Because $F_{pp}(t, x, p) > 0$ according to condition c) at the beginning of this section all the conditions for Theorem 1.4.2 are satisfied. For all $t_1$ and $t_2 \in \mathbb{R}$, $\overline{\gamma}$ is a minimal in

$$\Lambda(t_1, t_2) := \{\gamma : t \mapsto x(t) \mid x \in \text{Lip}(t_1, t_2), \ x(t_1) = \overline{x}(t_1), x(t_2) = \overline{x}(t_2)\}.$$

Let $\phi$ be an arbitrary element in $\text{Lip}_{comp}(\mathbb{R})$ and let $\tilde{\gamma}$ be given as $\tilde{x}(t) = \overline{x}(t) + \phi(t)$. Because $\phi$ has compact support there exists $T > 0$ so that $\tilde{\gamma} \in \Lambda(-T, T)$. Therefore, one has

$$\int_{\mathbb{R}} F(t, \tilde{x}, \dot{\tilde{x}}) - F(t, \overline{x}, \dot{\overline{x}}) \, dt \ = \ \int_{-T}^{T} F(t, \tilde{x}, \dot{\tilde{x}}) - F(t, \overline{x}, \dot{\overline{x}}) \, dt$$

$$= \ \int_{-T}^{T} E(t, \overline{x}, \dot{\tilde{x}}, \psi(t, \overline{x})) \, dt \geq 0 \, ,$$

where $E$ is the Weierstrass E-function. This means that $\tilde{\gamma}$ is a global minimal.

If $x(t)$ is an extremal solution to the extremal field, then also $y(t) = x(t + j) - k$ is an extremal solution, because $\psi$ is periodic in $t$ and $x$. If $x$ and $y$ have a self intersection, $x \equiv y$ follows by the uniqueness theorem for ordinary differential equations and $x, y$ satisfy the same differential equation

$$\dot{x} = \psi(t, x), \dot{y} = \psi(t, y) .$$

We have seen that every extremal solution in one extremal field is a global minimal. What about global minimals without an extremal field. Do they still exist? In the special case of the geodesic flow on the two dimensional torus, there exists only one metric for which all solutions are minimals. This is a theorem of Eberhard Hopf [16] which we cite here without proof.

---

**Theorem 2.1.2.** *(Hopf) If all geodesics on the torus are global minimals, then the torus is flat: the Gaussian curvature is zero.*

---

The relation of extremal fields with minimal geodesics will be treated later again, where we will also see that in general, global extremal fields do not need to exist. According to Theorem 1.5.1 an extremal field $\psi$ can be represented by a function $\psi = H(t, x, g_x)$, where $g(t, x)$ satisfies the **Hamilton–Jacobi equations**

$$g_t + H(t, x, g_x) = 0, \ g_x \in C^1(\mathbb{T}^2) .$$

The existence of a function $g$ on $\mathbb{T}^2$ solving the Hamilton–Jacobi equations globally is equivalent to the existence of a global extremal field. While it is well known how to solve the Hamilton–Jacobi equations locally, we deal here with a global problem and periodic boundary conditions. The theorem of Hopf shows that this problem can not be solved in general.

We will see that the problem has solutions if one widens the class of solutions. These will form **weak solutions** in some sense. The minimals will lead to weak solutions of the Hamilton–Jacobi equations.

## 2.2   An existence theorem

The aim of this section is to prove the existence and regularity of minimals with given boundary values or with periodic boundary conditions within a function class which is bigger then the function class considered so far. We will use here the assumptions (2.1) and (2.2) on quadratic growth.

Let $W^{1,2}[t_1, t_2]$ denote the Hilbert space obtained by closing $C^1[t_1, t_2]$ with respect to the norm

$$||x||^2 = \int_{t_1}^{t_2} (x^2 + \dot{x}^2) \, dt .$$

One calls it a **Sobolev space**. It contains $\text{Lip}[t_1, t_2]$, the space of Lipschitz continuous functions which is also denoted by $W^{1,\infty}$.

Analogously as we have dealt with variational problems in $\Gamma$ and $\Lambda$ we search now in

$$\Xi := \{\gamma : t \mapsto x(t) \in \mathbb{T}^2 \mid x \in W^{1,2}[t_1, t_2], x(t_1) = a, x(t_2) = b \}$$

for extremal solutions to the functional

$$I(\gamma) = \int_{t_1}^{t_2} F(t, x, \dot{x}) \, dt \ .$$

The set $\Xi$ is not a linear space. But with

$$x_0(t) = \frac{a(t_2 - t) + b(t - t_1)}{t_2 - t_1} \ ,$$

$\Xi = x_0 + \Xi_0$, where

$$\Xi_0 = \{\gamma : t \mapsto x(t) \in \mathbb{T}^2 \mid x \in W^{1,2}[t_1, t_2], x(t_1) = 0, x(t_2) = 0 \}$$

is a linear space.

---

**Theorem 2.2.1.** *It follows from conditions* (2.1) *to* (2.2) *that there exists a minimal* $\gamma^* : t \mapsto x^*(t)$ *in* $\Xi$. *Furthermore* $x^* \in C^2[t_1, t_2]$ *and* $x^*$ *satisfies the Euler equations.*

---

The proof is based on a basic principle: a lower semi-continuous function which is bounded from below takes a minimum on a compact topological space.

*Proof.* 1) $I$ **is bounded from below:**

$$\mu = \inf\{I(\gamma) \mid \gamma \in \Xi\} > -\infty \ .$$

From $\delta < F_{pp} < \delta^{-1}$ we obtain by integration: there exists $c$ with

$$\frac{\delta}{4}p^2 - c \leq F(t, x, p) \leq \delta^{-1}p^2 + c \ ,$$

so that for every $\gamma \in \Xi$,

$$I(\gamma) = \int_{t_1}^{t_2} F(t, x, \dot{x}) \, dt \geq \frac{\delta}{4} \int_{t_1}^{t_2} \dot{x}^2 \, dt - c(t_2 - t_2) \geq -c(t_2 - t_2) > -\infty \ .$$

This is called **coercivity**. Denote by $\mu$ the just obtained finite infimum of $I$.

2) **The closure of the set**

$$K := \{\gamma \in \Xi \mid I(\gamma) \leq \mu + 1 \}$$

**(using the topology given by the norm) is weakly compact.**

Given $\gamma \in K$. From

$$\mu + 1 \geq I(\gamma) \geq \frac{\delta}{4} \int_{t_1}^{t_2} \dot{x}^2 \, dt - c(t_2 - t_1)$$

follows

$$\int_{t_1}^{t_2} \dot{x}^2 \, dt \leq \frac{4}{\delta}(\mu + 1 + c(t_2 - t_1)) =: M_1 \, ,$$

and with $|x(t)| \leq a + \int_{t_1}^{t_2} \dot{x}(t) \, dt \leq a + [\int_{t_1}^{t_2} \dot{x}^2 \, dt(t_2 - t_1)]^{1/2}$ we get

$$\int_{t_1}^{t_2} x^2 \, dt \leq (t_2 - t_1)(a + [\frac{4}{\delta}(\mu + 1)(t_2 - t_1)]^{1/2})^2 =: M_2 \, .$$

Both together lead to

$$||\gamma||^2 = \int_{t_1}^{t_2} (\dot{x}^2 + x^2) \, dt \leq M_1 + M_2 \, .$$

This means that the set $K$ is bounded. Therefore its strong closure is bounded too. Because a bounded and closed set is weakly compact in $\Xi$, the closure of $K$ is weakly compact.

(It is an exercise to give a direct proof of this step using the theorem of Arzela–Ascoli.)

3) $I$ is lower semi-continuous in the weak topology.
We have to show that $I(\gamma) \leq \liminf_{n \to \infty} I(\gamma_n)$ if $\gamma_n \to_w \gamma$. (The symbol $\to_w$ denotes the convergence in the weak topology.)

a) The function $p \mapsto F(t, x, p)$ is convex:

$$F(t, x, p) - F(t, x, q) \geq F_p(t, x, q)(p - q) \, .$$

*Proof.* This is equivalent to $E(t, x, p, q) \geq 0$, an inequality which we have seen in the proof of Theorem 1.4.2.

b) If $x_n \to_w x$, then $\int_{t_1}^{t_2} \phi[\dot{x}_n - \dot{x}] \, dt \to 0$ for $\phi \in L^2[t_1, t_2]$.

*Proof.* The claim is clear for $\phi \in C^1$ by partial integration. Because $C^1$ is dense in $L^2$, we can for an arbitrary $\phi \in L^2$ and $\epsilon > 0$ find an element $\tilde{\phi} \in C^1$ so that $||\phi - \tilde{\phi}||_{L^2} < \epsilon$. We have then

$$\left| \int_{t_1}^{t_2} \phi(\dot{x}_n - \dot{x}) \, dt \right| \leq \left| \int_{t_1}^{t_2} \tilde{\phi}(\dot{x}_n - \dot{x}) \, dt \right| + 2\epsilon M_1 \, ,$$

and therefore

$$\limsup_{n\to\infty} \left| \int_{t_1}^{t_2} \phi(\dot{x}_n - \dot{x}) \, dt \right| \le 2\epsilon M_1 \ .$$

c) If $x_n \to_w x$, then $\int_{t_1}^{t_2} \phi[x_n - x] \, dt \to 0$ for $\phi \in L^2[t_1, t_2]$.

*Proof.* $x_n \to_w x$ implies that $x_n$ converges uniformly to $x$.

$\int_{t_1}^{t_2} \dot{x}_n^2 \, dt \le M_1$ implies that $|x_n(t) - x_n(s)| \le M_1(|t - s|)^{1/2}$ and $x_n(t) \le a + M(t - t_1)$. Therefore, $\{x_n \mid n \in \mathbb{N}\}$ is an equicontinuous family of uniformly bounded functions. According to Arzela–Ascoli, there exists a subsequence of $x_n$ which converges uniformly. Because $x_n \to_w x$, we must have $x$ as the limit. From $||x_n - x||_{L^\infty} \to 0$ follows with Hölders inequality that

$$\left| \int_{t_1}^{t_2} \phi[x_n - x] \, dt \right| \le \int_{t_1}^{t_2} |\phi| \, dt, \ ||x_n - x||_{L^\infty} \to 0 \ .$$

Using a),b) and c), we can now prove the claim:

$$
\begin{aligned}
I(\gamma_n) - I(\gamma) &= \int_{t_1}^{t_2} F(t, x, \dot{x}) - F(t, x_n, \dot{x}_n) \\
&\quad - F(t, x, \dot{x}_n) + F(t, x, \dot{x}_n) - F(t, x, \dot{x}) \, dt \\
&\ge \int_{t_1}^{t_2} F_x(t, \tilde{x}, \dot{x}_n)(x_n - x) \, dt \\
&\quad + \int_{t_1}^{t_2} F_p(t, x, \dot{\tilde{x}})(\dot{x}_n - \dot{x}) \, dt =: D_n \ .
\end{aligned}
$$

In that case, $\tilde{x}(t)$ is in the interval $[x_n(t), x(t)]$ and $\dot{\tilde{x}}$ is in the interval $[\dot{x}_n(t), \dot{x}(t)]$. For the inequality, we had used a). Since $F_x$ is in $L^1$ (because $|F_x| \le c(1+\dot{x}^2) \in L^1$), and $F_p$ is in $L^1$ (because $|F_p| \le c(1 + |\dot{x}|) \in L^2 \subset L^1$), we conclude with b) and c) that $D_n$ converges to 0 for $n \to \infty$. This finishes the proof:

$$\liminf_{n\to\infty}(I(\gamma_n) - I(\gamma)) \ge 0 \ .$$

**4) Existence of the minimals.**
The existence of minimals is accomplished from 1) to 3) and the fact that a lower semi-continuous function which is bounded from below takes a minimum on a compact space.

**5) Regularity of the minimals**.
Let $\gamma^* : t \mapsto x^*(t)$ be a minimal element in $\Xi$ from which we had proven existence in 4). For all $\phi : t \mapsto y(t)$, $\phi \in \Xi$

$$I(\gamma + \epsilon\phi) \ge I(\gamma^*) \ .$$

This means that the first variation must disappear if it exists.

Claim. The first variation $\lim_{\epsilon \to 0}(I(\gamma^* + \epsilon\phi) - I(\gamma^*))/\epsilon$ exists.

$$[I(\gamma^* + \epsilon\phi) - I(\gamma^*)]/\epsilon = \int_{t_1}^{t_2} [F(t, x^* + \epsilon y, \dot{x}^* + \epsilon\dot{y}) - F(t, x^*, \dot{x}^*)] \, dt/\epsilon$$

$$= \int_{t_1}^{t_2} [\lambda(t, \epsilon)\dot{y} + \mu(t, \epsilon)y] \, dt$$

with

$$\lambda(t, \epsilon) = \int_0^1 F_p(t, x^*, \dot{x}^* + \theta\epsilon\dot{y}) \, d\theta$$

$$\mu(t, \epsilon) = \int_0^1 F_x(t, x^* + \theta\epsilon y, \dot{x}^*) \, d\theta .$$

These estimates become for $\epsilon < 1$ and $\theta_0 \in [0, 1]$:

$$|\lambda(t, \epsilon)| \leq c(1 + |\dot{x}^* + \epsilon\theta_0\dot{y}|) \leq c(1 + |\dot{x}^*| + |\dot{y}|) ,$$
$$|\mu(t, \epsilon)| \leq c(1 + (\dot{x}^*)^2 + \dot{y}^2) .$$

According to the Lebesgue theorem, both $\lambda(t, \epsilon)\dot{y}$ and $\mu(t, \epsilon)y$ are in $L^1[t_1, t_2]$ because the majorants $c(1 + |\dot{x}^*| + |\dot{y}|)\dot{y}$ and $c(1 + (\dot{x}^*)^2 + \dot{y}^2)y$ are Lebesgue integrable. With the convergence theorem of Lebesgue follows the existence of $\lim_{\epsilon \to 0}[I(\gamma^* + \epsilon\phi) - I(\gamma^*)]/\epsilon = 0$ so that

$$\lim_{\epsilon \to 0}[I(\gamma + \epsilon\phi) - I(\gamma)]/\epsilon = \int_{t_1}^{t_2} F_p(t, x^*\dot{x}^*)\dot{y} + F_x(t, x^*, \dot{x}^*)y \, dt$$

$$= \int_{t_1}^{t_2} \left( F_p(t, x^*\dot{x}^*) - \int_{t_1}^s F_x(s, x^*\dot{x}^*) \, ds + c \right) \dot{y} \, dt$$

$$= 0 .$$

This means that

$$F_p(t, x^*, \dot{x}^*) = \int_{t_1}^t F_x(s, x^*, \dot{x}^*) \, ds + c$$

is absolutely continuous. From $F_{pp} > 0$ and the implicit function theorem we find $\dot{x}^* \in C^0$ and $x^* \in C^1$. From the integrated Euler equations we get $F_p \in C^1$. Again applying the implicit function theorem gives $\dot{x}^* \in C^1$ from which $x^* \in C^2$ is obtained.                                                                        $\square$

In the second part of this section we will formulate the corresponding theorem on the existence of periodic minimals.

**Definition.** A curve $\gamma : t \mapsto x(t)$ is **periodic of type** $(q, p)$ for $(q, p) \in \mathbb{Z}^2, q \neq 0$, if $x(t + q) - p \equiv x(t)$.

Define for $q \neq 0$,

$$\Xi_{p,q} = \{ \gamma : t \mapsto x(t) = \frac{p}{q} t + \xi(t) \mid \xi \in W^{1,2}[t_1, t_2], \ \xi(t + q) = \xi(t) \}$$

with the vector space operations

$$\rho \gamma_1 \quad : \quad \frac{p}{q} t + \rho \xi_1(t) \, ,$$

$$\gamma_1 + \gamma_2 \quad : \quad \frac{p}{q} t + \xi_1(t) + \xi_2(t) \, ,$$

if $\gamma_j : t \mapsto \frac{p}{q} t + \xi_j(t)$. The dot product

$$(\gamma_1, \gamma_2) = \int_0^q \xi_1 \xi_2 + \dot{\xi}_1 \dot{\xi}_2 \ dt$$

makes $\Xi_{p,q}$ a Hilbert space.

**Definition.** A minimal of the functional

$$I(\gamma) = \int_0^q F(t, x, \dot{x}) \ dt$$

is called a **periodic minimal of type** $(q, p)$. We write $\mathcal{M}(q, p)$ for the set of periodic minimals of type $(q, p)$.

We will sometimes also abbreviate $x \in \mathcal{M}(q, p)$ if $\gamma \in \mathcal{M}(q, p)$ is given by $\gamma : t \mapsto x(t)$.

---

**Theorem 2.2.2.** *For every $(q, p) \in \mathbb{Z}^2$ with $q \neq 0$, there exists an element $\gamma^* \in \mathcal{M}(q, p)$ with $\gamma : t \mapsto x^*(t)$ so that $x^* \in C^2(\mathbb{R})$ satisfies the Euler equations.*

---

The proof of Theorem 2.2.2 follows the same lines as the proof of Theorem 2.2.1.

**Remark on the necessity of the quadratic growth.**
The assumptions of quadratic growth (2.1)–(2.2) could be weakened. For the existence theorem it would suffice to assume superlinear growth. A classical theorem of **Tonelli** guarantees the existence of absolutely continuous minimals under the assumption that $F_{pp} \geq 0$ and

$$F(t, x, p) \geq \phi(p) := \lim_{|p| \to \infty} \frac{\phi(p)}{|p|} = \infty \, .$$

On the other hand, such an existence theorem no longer holds if $F$ has only linear growth in $p$. One can show for example that

$$F(x, p) = \sqrt{1 + p^2} + x^2 p^2$$

with boundary conditions

$$x(-1) = -a, x(1) = a$$

has no minimal for sufficiently large $a$, even though in this example, $F_{pp} > 0$ has only linear growth at $x = 0$. As a reference to the theorem of Tonelli and the above example see [9].

We also give an example without global minimals, where $F(t, x, p)$ is periodic in $t$ and $x$: let

$$F(t, x, p) = a(t, x)\sqrt{1 + p^2}$$

with $a(t, x) = 1 + b(t^2 + x^2)$ for $|t|, |x| \leq 1$. If $b = b(\lambda) \geq 0$, there exists a $C^\infty$-function, which vanishes identically outside the interval $[0.1, 0.2]$. We take $a(t, x)$ with period 1 in $t$ and periodically continue $a$ in $x$ to get a function on $\mathbb{R}^2$. Then, $a(t, x) \geq 1$ for all $t, x \in \mathbb{R}$ and the variational problem is

$$\int F(t, x, \dot{x})\, dt = \int a(t, x)\, ds\ ,$$

where $ds = \sqrt{1 + \dot{x}^2}\, dt$.

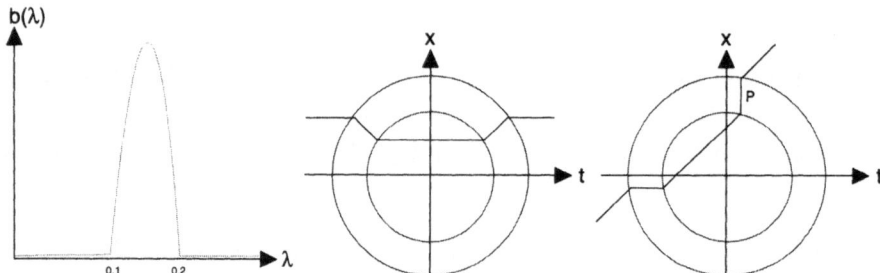

We consider a unique minimal segment, which is contained in the disc $t^2 + x^2 \leq 1/4$ and which is not a straight line. Now we use the rotational symmetry of the problem and turn the segment in such a way that it can be represented as a graph $x = x(t)$, but so that $\dot{x}(\tau) = \infty$ for a point $P = (\tau, x(\tau))$.

Because this segment is a unique minimal for the corresponding boundary condition, it must have a singularity at $t = \tau$. The condition of quadratic growth excludes such a singular behavior.

## 2.3   Properties of global minimals

In this section we derive properties for global minimals, which will allow us to construct them in the next section. Throughout this section we always assume that the dimension is 1.

**Definition.** Denote by $\mathcal{M}$ the set of **global minimals**. Given $x$ and $y$ in $\mathcal{M}$ we write

$$x \leq y, \quad \text{if } x(t) \leq y(t), \; \forall t \,,$$
$$x < y, \quad \text{if } x(t) < y(t), \; \forall t \,,$$
$$x = y, \quad \text{if } x(t) = y(t), \; \forall t \,.$$

---

**Theorem 2.3.1.** *a) Two different global minimals $x$ and $y$ in $\mathcal{M}$ do intersect at most once.*
*b) If $x \leq y$, then $x = y$ or $x < y$.*
*c) If $\lim_{t \to \infty} |x(t) - y(t)| + |\dot{x}(t) - \dot{y}(t)| = 0$ and $\sup_{t>0}(|x(t)| + |y(t)|) \leq M < \infty$ for $x < y$ or $x > y$.*
*d) Two different periodic minimals of type $(q, p)$ do not intersect.*

---

*Proof.* a) Let $x$ and $y$ be two global minimals which intersect twice in the interval $[t_1, t_2]$. The same argument as in the proof of Theorem 1.3.4 with $\Xi$ instead of the function space $\Lambda$ leads to a contradiction.

b) Assume $x(t) = y(t)$ for some $t \in \mathbb{R}$. Then, $x \leq y$ implies $\dot{x}(t) = \dot{y}(t)$. The functions are differentiable and even $C^2$. According to the uniqueness theorem for ordinary differential equations applied to the Euler equations $\ddot{x} F_{pp} + \dot{x} F_{xp} + F_{tp} = F_x$, we must have $x = y$.

c) Assume the claim is wrong and that there exists a time $t \in \mathbb{R}$ with $x(t) = y(t)$. Claim (*):

$$\lim_{T \to \infty} \left| \int_\tau^T F(t, x, \dot{x}) \, dt - \int_\tau^T F(t, y, \dot{y}) \, dt \right| = 0 \,.$$

Proof. We can construct $z$ as follows:

$$z(t) = \begin{cases} y(t) & , \; t \in [\tau, T-1] \,, \\ x(t) - (t-T)(y(t) - x(t)) & , \; t \in [T-1, T] \end{cases}$$

Because of the minimality of $x$ we have

$$\int_\tau^T F(t, x, \dot{x}) \, dt \quad \leq \quad \int_\tau^T F(t, z, \dot{z}) \, dt$$

$$= \quad \int_\tau^T F(t, y, \dot{y}) \, dt + \int_\tau^T F(t, z, \dot{z}) - F(t, y, \dot{y}) \, dt$$

$$= \quad \int_\tau^T F(t, y, \dot{y}) \, dt + \int_{T-1}^T F(t, z, \dot{z}) - F(t, y, \dot{y}) \, dt \,.$$

For $t \in [T-1, T]$ the point $(x(t), \dot{x}(t))$ is by assumption contained in the compact set $\mathbb{T}^2 \times [-M, M]$. The set $\Pi = [T-1, T] \times \mathbb{T}^2 \times [-M, M]$ is compact in the phase space $\Omega \times \mathbb{R}$. Now

$$\int_{T-1}^{T} F(t, z, \dot{z}) - F(t, y, \dot{y}) \, dt$$
$$\leq \max_{(t,u,v) \in \Pi} \{ F_x(t, u, v) |z(t) - y(t)| + F_p(t, u, v) |\dot{z}(t) - \dot{y}(t)| \}$$
$$\to \quad 0$$

for $T \to \infty$ because of the assumptions on $|y(t) - x(t)|$ and $|\dot{y}(t) - \dot{x}(t)|$. One has $(z(t) - y(t)) = (x(t) - y(t))(1 + t - T)$ for $t \in [T-1, T]$ finishing the proof of claim (*). On the other hand, the minimals $x(t)$ and $y(t)$ have to intersect transversally at a point $t = \tau$. Otherwise they would coincide according to the uniqueness theorem of differential equations. This means that there exists $\epsilon < 0$ so that the path $t \mapsto x(t)$ is not minimal on $[\tau - \epsilon, T]$ for large enough $T$. Therefore

- On the interval $[\tau - \epsilon, \tau + \epsilon]$ the action can be decreased by a fixed positive value if a minimal $C^2$-path $q(t)$ is chosen from $x(t - \epsilon)$ to $y(t + \epsilon)$ instead of continuing $x(t)$ and $y(t)$ along $[\tau - \epsilon, \tau + \epsilon]$ and going around corners.

- According to claim (*) the difference of the actions of $x(t)$ and $y(t)$ on the interval $[\tau, T]$ can be made arbitrarily small if $T$ goes to $\infty$.

- The path

$$t \mapsto \begin{cases} q(t) & , t \in [\tau - \epsilon, \tau + \epsilon] \, , \\ z(t) & , t \in [\tau - \epsilon, T] \, , \end{cases}$$

  has therefore for large enough $T$ a smaller action than $x(t)$. This contradicts the assumption that $x(t)$ is a global minimal.

d) Instead of looking for a minimum of the functional

$$I(\gamma) = \int_0^q F(t, x, \dot{x}) \, dt$$

in $\Xi_{q,p}$, we can seek a minimum of

$$I_\epsilon(\gamma) = \int_\epsilon^{q+\epsilon} F(t, x, \dot{x}) \, dt$$

because both functionals coincide on $\Xi_{q,p}$. If $\gamma$ has two roots in $(0, q]$, we can find $\epsilon > 0$, so that $\gamma$ has two roots in $(\epsilon, q + \epsilon)$. Therefore, $I_\epsilon(\gamma)$ can not be minimal, by the same argument as in a), and therefore also not on $I(\gamma)$. The function $\gamma$ has therefore at most one root in $(0, q]$. According to Theorem 2.3.2 a) below (which uses in the proof only part a) of this theorem) $\gamma$ has therefore also at most one root in $(0, Nq]$, but is periodic with period $q$. $\qquad\square$

**Theorem 2.3.2.** *For all $N \in \mathbb{N}, (q, p) \in \mathbb{Z}, q \neq 0$ one has:*
*a) $\gamma \in \mathcal{M}(q, p)$ if and only if $\gamma \in \mathcal{M}(Nq, Np)$.*
*b) The class $\mathcal{M}(q, p)$ is characterized by $p/q \in \mathbb{Q}$.*
*c) $\mathcal{M}(q, p) \subset \mathcal{M}$.*

*Proof.* a) (i) Let $\gamma \in \mathcal{M}(Nq, Np)$ be defined as

$$\gamma : x(t) = \frac{p}{q} t + \xi(t)$$

with $\xi(t + Nq) = \xi(t)$. We claim that $\gamma \in \mathcal{M}(q, p)$. Put $y(t) = x(t + q) = \frac{p}{q} t + \eta(t)$ with $\eta(t) = x(t + q)$. Since $x(t) - y(t) = x(t) - \eta(t) = x(t) - x(t + q)$ has period $Nq$ and

$$\int_0^{Nq} (x - y) \, dt = \int_0^{Nq} (\xi(t) - \xi(t + q)) \, dt = 0 \, ,$$

$x(t) - y(t)$ disappears by the intermediate value theorem at two places in $(0, Nq)$ at least. Theorem 2.3.1 a) implies $x = y$ and

$$I(\gamma)|_0^{Nq} = \int_0^{Nq} F(t, x, \dot{x}) \, dt = N \int_0^q F(t, x, \dot{x}) \, dt = NI(\gamma)|_0^q \, .$$

Because $\Xi_{Nq, Np} \supset \Xi_{q, p}$,

$$\inf_{\eta \in \Xi_{q, p}} I(\eta)|_0^q \geq \inf_{\eta \in \Xi_{Nq, Np}} N^{-1} I(\eta)|_0^{Nq} = N^{-1} I(\gamma)|_0^{Nq} = I(\gamma)|_0^q$$

proving $\gamma \in \mathcal{M}(q, p)$.

(ii) On the other hand, if $\gamma \in \mathcal{M}(q, p)$ is given, we show that $\gamma \in \mathcal{M}(Nq, Np)$. The function $\gamma$ is also an element of $\Xi_{Nq, Np}$. According to the existence Theorem 2.2.2 in the last section, there exists a minimal element $\zeta \in \mathcal{M}(Nq, Np)$ for which we have

$$NI(\gamma)_0^q = I(\gamma)|_0^{Nq} > I(\zeta)|_0^{Nq} \, .$$

From (i), we conclude that $\zeta \in \mathcal{M}(q, p)$ and that

$$NI(\gamma)|_0^q \geq NI(\zeta)|_0^q \, .$$

Because $\gamma \in \mathcal{M}(q, p)$ we also have $NI(\gamma)|_0^q \leq NI(\zeta)|_0^q$ and therefore

$$NI(\gamma)|_0^q = NI(\zeta)|_0^q$$

and finally

$$I(\gamma)|_0^{Nq} = I(\zeta)|_0^{Nq}$$

which means that $\gamma \in \mathcal{M}(Nq, Np)$.

b) follows immediately from a).

c) Let $\gamma \in \mathcal{M}(q, p)$. We have to show that for $\phi \in \text{Lip}_{comp}(\mathbb{R})$,

$$\int_{\mathbb{R}} F(t, x + \phi, \dot{x} + \dot{\phi}) - F(t, x, \dot{x}) \, dt \geq 0 .$$

Choose $N$ so big that the support of $\phi$ is contained in the interval $[-Nq, Nq]$. Call $\tilde{\phi}$ the $2Nq$-periodic continuation of $\phi$. Since $\gamma \in \mathcal{M}(q, p)$,

$$\int_{\mathbb{R}} F(t, x + \phi, \dot{x} + \dot{\phi}) - F(t, x, \dot{x}) \, dt \;=\; \int_{-Nq}^{Nq} F(t, x + \phi, \dot{x} + \dot{\phi}) - F(t, x, \dot{x}) \, dt$$

$$=\; \int_{-Nq}^{Nq} F(t, x + \tilde{\phi}, \dot{x} + \dot{\tilde{\phi}}) - F(t, x, \dot{x}) \, dt$$

$$\geq\; 0 . \qquad \square$$

Theorem 7.2 can be summarized as follows: a periodic minimal of type (q,p) is globally minimal and characterized by a rational number $p/q$. We write therefore $\mathcal{M}(p/q)$ instead of $\mathcal{M}(q, p)$.

---

**Theorem 2.3.3.** *Global minimals have no self intersections on* $\mathbb{T}^2$.

---

**Definition.** Denote by $\mathcal{M}[0, T]$ the set of minimals on the interval $[0, T]$.

The proof of Theorem 2.3.3 needs estimates for elements in $\mathcal{M}[0, T]$:

---

**Lemma 2.3.4.** *Let* $\gamma \in \mathcal{M}[0, T], \gamma : t \mapsto x(t)$ *and* $A > T > 1$, *so that* $|x(T) - x(0)| \leq A$. *There are constants* $c_0, c_1, c_2$ *which only depend on* $F$, *so that for all* $t \in [0, T]$:

$$a) \quad |x(t) - x(0)| \leq C_0(A) = c_0 A , \qquad\qquad (2.6)$$

$$b) \quad |\dot{x}(t)| \leq C_1(A) = c_1 A^2 T^{-1} , \qquad\qquad (2.7)$$

$$c) \quad |\ddot{x}(t)| \leq C_2(A) = c_2 A^4 T^{-2} . \qquad\qquad (2.8)$$

---

*Proof.* $\eta : t \mapsto y(t) \in \mathcal{M}[0, T]$. From $\delta \leq F_{pp} \leq \delta^{-1}$ we get by integration (compare Theorem 2.2.1):

$$-a_1 + \frac{\delta}{4}\dot{x}^2 \leq F \;\leq\; \delta^{-1}\dot{x}^2 + a_1 ,$$

$$-a_1 + \frac{\delta}{4}\dot{y}^2 \leq F \;\leq\; \delta^{-1}\dot{y}^2 + a_1 .$$

Because of the minimality of $\gamma$ the inequality $I(\gamma) \leq I(\eta)$ holds. With $y = T^{-1}(x(0)(T-t) + x(T)t)$ we have

$$
\begin{aligned}
-a_1 T + \frac{\delta}{4} \int_0^T \dot{x}^2 \, dt &\leq \int_0^T F(t, x, \dot{x}) \, dt \\
&\leq \int_0^T F(t, y, \dot{y}) \, dt \\
&\leq \delta^{-1} \int_0^T \dot{y}^2 \, dt + a_1 T \\
&\leq \delta^{-1}[x(T) - x(0)]^2 T^{-1} + a_1 T \\
&\leq \delta^{-1} A^2 T^{-1} + a_1 T .
\end{aligned}
$$

We conclude that

$$
\int_0^T \dot{x}^2 \, dt \leq 4\delta^{-2} A^2 T^{-1} + 8a_1 T \delta^{-1} \leq a_2 A^2 T^{-1} .
$$

Now the proof of claim a) can be finished:

$$
\begin{aligned}
|x(t) - x(0)| &= \left| \int_0^t 1 \cdot \dot{x} \, ds \right| \\
&\leq \sqrt{t} \left[ \int_0^t \dot{x}^2 \, ds \right]^{1/2} \\
&\leq \sqrt{T} [a_2 A^2 T^{-1}]^{1/2} = c_0 A .
\end{aligned}
$$

Because $\gamma$ is in $\mathcal{M}[0, T]$, the function $x(t)$ satisfies the Euler equations $\frac{d}{dt} F_p = F_x$, which are

$$
\ddot{x} F_{pp} + \dot{x} F_{xp} + F_{tp} = F_x .
$$

With $F_{pp} \geq \delta$, $|F_x| \leq c(1 + \dot{x}^2)$, $|F_{xp}| \leq c(1 + |\dot{X}|)$ and $|F_{tp}| \leq c(1 + |\dot{x}|)$, we can estimate $\ddot{x}$ as follows: there exists a constant $a_3$ with

$$
|\ddot{x}| \leq a_3(1 + \dot{x}^2)
$$

and therefore also b) is proven: for all $t, s \in [0, T]$ one has

$$
|\dot{x}(t) - \dot{x}(s)| = \left| \int_s^t \ddot{x} \, dt \right| \leq a_3 \int_0^T (1 + \dot{x}^2) \, dt \leq a_3[T + a_2 A^2 T^{-1}] \leq a_4 4 A^2 T^{-1} .
$$

There exists $s \in [0, T]$ with $|\dot{x}(s)| = |[x(T) - x(0)]T^{-1}| \leq A T^{-1}$ so that

$$
|\dot{x}(t)| \leq A T^{-1} + a_4 A^2 T^{-1} \leq c_1 A^2 T^{-1} .
$$

c) is done by noting:

$$
|\ddot{x}(t)| < a_3(1 + \dot{x}^2) \leq a_3[1 + (c_1 A T^{-1})^2] \leq c_2 A^4 T^{-2} . \qquad \square
$$

We turn now to the proof of Theorem 2.3.3 which said that global minimals have no self intersections.

*Proof.* Assume that there exists $\gamma \in \mathcal{M}$ with a self intersection. This means, there exists $(q, p) \in \mathbb{Z}^2$, $q \neq 0$ and $\tau \in \mathbb{R}$ with

$$x(\tau + q) - p = x(\tau) \ .$$

Without loss of generality we can assume $\tau = 0$. Writing $x(t) = \frac{p}{q}t + \xi(t)$ one has

$$x(\tau + q) - p = \frac{p}{q}t + \xi(t + q) \ .$$

Because there is at most one intersection of $x(t)$ and $x(t + q) - p$, we have

$$\left. \begin{array}{ll} x(t+q) - p - x(t) > 0, & t > 0 \\ x(t+q) - p - x(t) < 0, & t < 0 \end{array} \right\} \text{ which means } \left\{ \begin{array}{ll} \xi(t+q) - \xi(t) > 0, & t > 0 \\ \xi(t+q) - \xi(t) < 0, & t < 0 \end{array} \right.$$

or

$$\left. \begin{array}{ll} x(t+q) - p - x(t) < 0, & t > 0 \\ x(t+q) - p - x(t) > 0, & t < 0 \end{array} \right\} \text{ which means } \left\{ \begin{array}{ll} \xi(t+q) - \xi(t) < 0, & t > 0 \\ \xi(t+q) - \xi(t) > 0, & t < 0 \end{array} \right. .$$

We can restrict ourselves without loss of generality to the first case. (Otherwise, replace $t$ by $-t$.) From

$$\left. \begin{array}{ll} \xi(t+q) - p - \xi(t) < 0, & t > 0 \\ \xi(t+q) - p - \xi(t) > 0, & t < 0 \end{array} \right\} \text{ which means } \left\{ \ \xi(t) - \xi(t - q) < 0, \ t < q \right.$$

follows that for every $n \in \mathbb{N}$,

$$\begin{array}{llll} \xi_n(t) & := & \xi(t + nq) > \xi_{n-1}(t), & t > 0 \ , \hspace{2em} (2.9) \\ \xi_n(t) & := & \xi(t - nq) > \xi_{-n+1}(t), & t < q \ . \hspace{2em} (2.10) \end{array}$$

Therefore, $\xi_n(t)$ is a monotonically increasing sequence for fixed $t > 0$ and $\xi_{-n}(t)$ is monotonically increasing for $t < q$ and $n \to \infty$ also. According to the existence theorem for periodic minimals in the last section, there exists a periodic minimal $\theta \in \mathcal{M}(q, p)$ $\theta : t \mapsto z(t), z(t) = \frac{p}{q}t + \zeta(t)$ with $\zeta(t) = \zeta(t + q)$. The requirement

$$z(0) < x(0) < z(0) + 2$$

can be achieved by a translation of $z$. We have therefore

$$\zeta(0) < \xi(0) = \xi(q) < \zeta(0) + 2 \ .$$

Because $\gamma$ and $\theta$ can not intersect two times in $[0, q]$, we have for $t \in [0, q]$

$$\begin{array}{llll} z(t) & < & x(t) < z(t) + 2 \ , \\ \zeta(t) & < & \xi(t) < \zeta(t) + 2 \ . \end{array}$$

Because $\zeta(t + nq) = \zeta(t)$ and $\xi_n(t) > \xi_{n-1}(t) > \xi(t)$, $\xi_{-n}(t) > \xi_{-n+1}(t)$ for $t \in [0, q]$, for all $n > 0$ and $t \in [0, q]$, either

$$\zeta(t) < \xi_n(t) < \zeta(t) + 2$$

or

$$\zeta(t) > \xi_n(t) > \zeta(t) + 2 \,.$$

If both estimates were wrong, there would exist $t', t'' \in [0, q]$ and $n', n'' > 0$ with

$$\begin{aligned}
\xi_{n'}(t') &= \zeta(t') + 2 \,, \\
\xi_{n''}(t'') &= \zeta(t'') + 2 \,,
\end{aligned}$$

which would lead to two intersections of $x(t)$ and $z(t)$ at $t = t' + n'$ and $t = t'' + n''$. Again we can restrict ourselves to the first case so that for all $t > 0$ the inequalities $\zeta(t) < \xi_n(t) < \xi_{n+1}(t) < \zeta(t) + 2$ hold for $t > 0$, where $\zeta(t)$ has period $q$. This means that there exists $\kappa(t)$, with $\xi_n(t) \to \kappa(t)$ for $n \to \infty$, pointwise for every $t > 0$. Because $\xi_{n+1}(t) = \xi(t + q) \to \kappa(t + q) = \kappa(t)$, this $\kappa$ has period $q$. If we can prove the three claims

$$\begin{aligned}
&i) && \exists M, \; |\dot{x}(t)| \le M, \; t > 0 \,, \\
&ii) && |x(t + q) - p - x(t)| \to 0, \; t \to \infty \,, \\
&iii) && |\dot{x}(t + q) - p - \dot{x}(t)| \to 0, \; t \to \infty \,,
\end{aligned}$$

we are finished by applying Theorem 2.3.1 c) to the global minimals given by $x(t)$ and $y(t) = x(t + q) - p$. The inequalities $x < y$ or $y < x$ mean that $\gamma$ can have no self intersections in contradiction to the assumption.

The claims i) to iii) follow in a similar way as in Lemma 2.3.4. They are equivalent to

$$\begin{aligned}
&i)' && \exists M, \; |\xi_n(t)| \le M, \; t \in [0, T] \,, \\
&ii)' && |\xi_{n+1}(t) - \xi_n(t)| \to 0, \; n \to \infty, t \in [0, q] \,, \\
&iii)' && |\dot{\xi}_{n+1}(t) - \dot{\xi}_n(t)| \to 0, \; n \to \infty, t \in [0, q] \,.
\end{aligned}$$

Claim i)' has already been proven by giving the periodic function $\kappa(t)$. With Lemma 2.3.4 we see that

$$\begin{aligned}
|\xi_n(t)| &\le C_1 \,, \\
|\dot{\xi}_n(t)| &\le C_2 \,.
\end{aligned}$$

This means that $\xi_n(t)$ and $\dot{\xi}_n(t)$ form an equicontinuous uniformly bounded sequence of functions. According to the theorem of Arzela–Ascoli, they converge uniformly. So, also (ii) and (iii) are proven. $\qquad \square$

As a corollary of Theorem 3.2.2, we see that if $\gamma \in \mathcal{M}$ and $\gamma$ is not periodic, an **order on** $\mathbb{Z}^2$ is defined by

$$(j, k) < (j', k') \text{ if } x(t + j) - k < x(t + j') - k', \forall t \,. \tag{2.11}$$

Any two allowed pairs $(j, k)$ and $(j', k')$ can be compared: $(j, k) < (j', k')$ or $(j, k) > (j', k')$.

## 2.4   A priori estimates and a compactness property

**Theorem 2.4.1.** *For a global minimal $\gamma \in \mathcal{M}, \gamma : t \mapsto x(t)$, the limit*

$$\alpha = \lim_{t \to \infty} \frac{x(t)}{t}$$

*exists.*

**Definition.** For $\gamma \in M$, the limiting value $\alpha = \lim_{t \to \infty} \frac{x(t)}{t}$ is called the **rotation number** or the **average slope** of $\gamma$.

The proof is based on the fact that the minimal $\gamma$ and its translates $T_{qp}\gamma : t \mapsto x(t + q) - p$ do not intersect.

*Proof.* **The first part of the proof.**

1) It is enough to show that the sequence $x(j)/j$ for $j \in \mathbb{Z}$, converges. According to Lemma 2.3.4 with $T = 1$ and $A = |x(j + 1) - x(j)| + 1$, for $t \in [j, j + 1]$ and $j > 0$ we have
$$|x(t) - x(j)| < c_0(|x(j + 1) - x(j)| + 1)$$
and

$$
\begin{aligned}
\left| \frac{x(t)}{t} - \frac{x(j)}{j} \right| &\leq \left| \frac{x(t) - x(j)}{t} + x(j)(\frac{1}{t} - \frac{1}{j}) \right| \\
&\leq \frac{|x(t) - x(j)|}{j} + \frac{|x(j)|}{j} \frac{(t - j)}{t} \\
&\leq \frac{|x(t) - x(j)|}{j} + \frac{|x(j)|}{j} \frac{1}{t} \\
&\leq c_0 \left| \frac{x(j + 1) - x(j)}{j} \right| + \frac{|x(j)|}{j} \frac{1}{t} \,.
\end{aligned}
$$

If we assume that $\alpha = \lim_{j \to \infty} x(j)/j$ exists, we have

$$\lim_{t \to \infty} \left| \frac{x(t)}{t} - \frac{x(j)}{j} \right| = 0 \,.$$

2) Because $x(t)$ has no self intersections, the map

$$f : S \to S, S = \{x(j) - k, (j, k) \in \mathbb{Z}^2\}, s = x(j) - k \mapsto f(s) = x(j+1) - k$$

is monotone and commutes with $s \mapsto s + 1$. This means

$$\begin{aligned} f(s) &< f(s'), s < s' \\ f(s+1) &= f(s) + 1 \, . \end{aligned}$$

In other words, $\hat{f}(s) := f(s) - s$ is periodic with period 1. $\qquad\square$

---

**Lemma 2.4.2.** $\forall s, s' \in S, \ |\hat{f}(s) - \hat{f}(s')| < 1.$

---

*Proof.* We assume that the claim is wrong and that there exists $s$ and $s' \in S$ such that

$$|\hat{f}(s) - \hat{f}(s')| \geq 1 \, .$$

We can assume without restricting generality that $\hat{f}(s) \geq \hat{f}(s') + 1$. We also can assume that $s < s' < s + 1$ by the periodicity of $\hat{f}$. We have therefore

$$f(s) - s - f(s') + s' \geq 1 \, . \tag{2.12}$$

The monotonicity of $f$ implies for $s < s' < s + 1$ that

$$f(s) < f(s') < f(s+1) \, .$$

From this, we get

$$f(s) + s' < f(s') + s + 1 \, . \tag{2.13}$$

Equation (2.12) contradicts Equation (2.13). $\qquad\square$

**Continuation of the proof.** The iterates of f,

$$f^m : x(j) - k \mapsto x(j+m) - k \, ,$$

exist for every $m \in \mathbb{Z}$ and $f^m$ has the same properties as $f$.

3) The numbers

$$\begin{aligned} b(f) &= \sup_{s \in S} \hat{f}(s) \, , \\ a(f) &= \inf_{s \in S} \hat{f}(s) \, , \end{aligned}$$

exist because of Lemma 2.3.4. Also

$$b(f) - a(f) \leq 1 \, .$$

In particular, both are finite because

$$b \leq 1 + (f(s_0) - s_0) < \infty, s_0 = x(0) .$$

$a$ and $b$ are subadditive:

$$b(f^{j+k}) \leq b(f^j) + b(f^k) ,$$
$$a(f^{j+k}) \leq a(f^j) + a(f^k) ,$$

because

$$\sup(f^{j+k}(s) - s) \leq \sup(f^j(f^k(s) - f^k(s))) + \sup(f^k(s) - s) \leq b(f^j) + b(f^k) .$$

It is well known that in this case

$$\lim_{j \to \infty} \frac{b(f^j)}{j} = \beta ,$$
$$\lim_{j \to \infty} \frac{a(f^j)}{j} = \alpha$$

exist. Because

$$0 \leq b(f^n) - a(f^n) \leq 1$$

holds, $\alpha = \beta$. The theorem is proven. $\qquad \square$

The result in Theorem 2.4.1 can be improved quantitatively. By the subadditivity of $a$ and $b$, one has:

$$a(f^k) \geq ka(f) ,$$
$$b(f^k) \leq kb(f) ,$$

and therefore

$$a(f) \leq \frac{a(f^m)}{m} \leq \frac{b(f^m)}{m} \leq b(f)$$

which gives with $m \to \infty$,

$$a(f) \leq \alpha \leq b(f) .$$

This means

$$-1 \leq \alpha(f) - b(f) \leq \hat{f}(s) - \alpha \leq b(f) - a(f) \leq 1 .$$

We have proven the following lemma:

---

**Lemma 2.4.3.** $|f(s) - s - \alpha| \leq 1, \forall s \in S .$

---

If Lemma 2.4.3 is applied to $f^m$, it gives

$$|f^m(s) - s - m\alpha| \leq 1, \forall s \in S .$$

This is an improvement of Theorem 2.4.1:

$$\left| \frac{f^m(s) - s}{m} - \alpha \right| \leq \frac{1}{m} .$$

Especially,

$$|x(m) - x(0) - m\alpha| \leq 1 .$$

---

**Theorem 2.4.4.** *If* $\gamma : t \mapsto x(t)$ *is a global minimal, then* $\forall t \in \mathbb{R}, \forall m \in \mathbb{Z}$,

$$|x(t + m) - x(t) - m\alpha| \leq 1 .$$

---

*Proof.* If instead of the function $F(t, x, \dot{x})$ the translated function $F(t + \tau, x, \dot{x})$ is taken, we get the same estimate as in Lemma 2.4.3 and analogously

$$|x(t + m) - x(t) - m\alpha| \leq 1 . \qquad \square$$

---

**Theorem 2.4.5.** *There is a constant c, which depends only on F (especially not on* $\gamma \in \mathcal{M}$ *nor on* $\alpha$*), so that for all* $t, t' \in \mathbb{R}$,

$$|x(t + t') - x(t) - \alpha t'| < c\sqrt{1 + \alpha^2} .$$

---

*Proof.* Choose $j \in \mathbb{Z}$ so that $j \leq t' \leq j + 1$.

Lemma 2.4.3 applied to $s = x(t + j)$ gives

$$|x(t + j + 1) - x(t + j)| = |f(s) - s| \leq |\alpha| + 1$$

which according to Lemma 2.3.4 with $T = 1$ and $A = 1 + |\alpha|$ gives

$$|x(t + t') - x(t + j)| \leq c_0(|\alpha| + 1) .$$

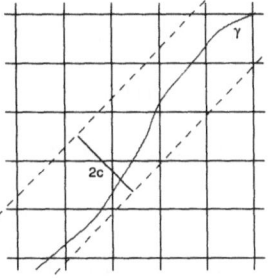

Using this and Theorem 2.4.4 we obtain Lemma 2.3.4.

$$
\begin{aligned}
|x(t + t') - x(t) - \alpha t'| &\leq |x(t + j) - x(t) - \alpha j| + |x(t + t') - x(t + j)| \\
&\quad + |\alpha|(t' - j)| \\
&\leq |1 + c_0(|\alpha| + 1) + |\alpha| = (c_0 + 1)(|\alpha| + 1) \\
&\leq 2(c_0 + 1)\sqrt{\alpha^2 + 1} \\
&=: c\sqrt{\alpha^2 + 1} . \qquad \square
\end{aligned}
$$

Theorem 2.4.5 has the following geometric interpretation: a global minimal $\gamma$ is contained in a strip of width $2c$. The width $2c$ is independent of $\gamma$ and $\alpha$!

From Theorem 2.4.1 follows that there exists a function $\alpha : \mathcal{M} \to \mathbb{R}, \gamma \mapsto \alpha(\gamma)$ which assigns to a global minimal its rotation number.

**Definition.** We define

$$\mathcal{M}_\alpha = \{\gamma \in \mathcal{M} \mid \alpha(\gamma) = \alpha\} \subset \mathcal{M}.$$

---

**Lemma 2.4.6.** *a)* $\mathcal{M} = \bigcup_{\alpha \in \mathbb{R}} = \mathcal{M}_\alpha$.
*b)* $\mathcal{M}_\alpha \cap \mathcal{M}_\beta = \emptyset, \alpha \neq \beta$.
*c)* $\mathcal{M}_{p/q} \supset \mathcal{M}(p/q) \neq \emptyset$.

---

*Proof.* a) and b) follow from Theorem 2.4.1.
c) $\mathcal{M}_{p/q} \supset \mathcal{M}(p/q)$ is obvious. The fact that $\mathcal{M}(p/q) \neq \emptyset$ was already proven in Theorem 2.2.1.                                                                                 □

---

**Theorem 2.4.7.** *Let* $\gamma \in \mathcal{M}_\alpha, \gamma : t \mapsto x(t), |\alpha| \leq A \geq 1$. *Then there exist constants* $d_0, d_1$ *and* $d_2$, *so that for all* $t, t_1, t_2 \in \mathbb{R}$:

$$
\begin{aligned}
a) \qquad & |x(t_1) - x(t_2) - \alpha(t_1 - t_2)| \leq D_0(A) := d_0 A, \\
b) \qquad & |\dot{x}(t)| \leq D_1(A) = d_1 A^2, \\
c) \qquad & |\ddot{x}(t)| \leq D_2(A) := d_2 A^4.
\end{aligned}
$$

---

*Proof.* Claim a) follows directly from Theorem 2.4.5:

$$|x(t_1) - x(t_2) - \alpha(t_1 - t_2)| \leq c\sqrt{1+\alpha^2} \leq \sqrt{2}c|\alpha| \leq \sqrt{2}A =: d_0 A.$$

b) From a), we get

$$|x(t+T) - x(t)| < |\alpha|T + d_0 A \leq A(T + d_0),$$

which give with Lemma 2.3.4 and with the choice $T = 1$,

$$|\dot{x}(t)| \leq c_1[A(T + d_0)]^2 T^{-1} = d_1 A^2.$$

c) Because

$$|\ddot{x}| \leq a_3(1 + |\dot{x}|^2) \leq a_3(1 + d_1^2 A^4) \leq 2a_3 d_1^2 A^4 = d_2 A^4$$

(compare Lemma 2.3.4 in the last section), also the third estimate is true.      □

**Remark**. Denzler [10] has given estimates of the form

$$D_1(A) = e^{d_1 A} .$$

The improvements in Theorem 2.4.7 use the minimality property and are likely not optimal. One expects

$$
\begin{aligned}
D_1(A) &= d_1 A , \\
D_2(A) &= d_2 A^2 ,
\end{aligned}
$$

an estimate holding for $F = (1 + \frac{1}{2}\sin(2\pi x))p^2$ because

$$E = (1 + \frac{1}{2}\sin(2\pi x))\dot{x}^2$$

is an integral of motion and $A$ is of the order $\sqrt{E}$.

**Definition**. We write $\mathcal{M}/\mathbb{Z}$ for the quotient space given by the equivalence relation $\sim$ on $\mathcal{M}$:

$$x \sim y \Leftrightarrow \exists k \in \mathbb{Z}, x(t) = y(t) + k .$$

In the same way we define the quotient $\mathcal{M}_\alpha/\mathbb{Z}$ on the subsets $\mathcal{M}_\alpha$.

**Definition**. The $C^1(\mathbb{R})$ topology on $C^1$-functions on $\mathbb{R}$ is defined as follows: $x_m(t) \to x(t), m \to \infty$ if for all compact $K \subset \mathbb{R}$, the sequence $x_m$ converges uniformly in the $C^1(K)$ topology. Analogously, for $r \geq 0$, the $C^r(\mathbb{R})$ topologies are defined. On the space of $C^1$-curves $\gamma : \mathbb{R} \to \Omega, t \mapsto x_m(t)$, the $C^1(\mathbb{R})$ topology is given in a natural way by $\gamma_n \to \gamma$ if $x_m \to x$ in $C^1(\mathbb{R})$.

---

**Lemma 2.4.8.** *$\alpha$ is continuous on $\mathcal{M}$, if we take the $C^0(\mathbb{R})$ topology on $\mathcal{M}$.*

---

*Proof.* We have to show that $x_m \to x$ implies $\alpha_m := \alpha(x_m) \to \alpha := \alpha(x)$. Because Theorem 2.4.7 gives $|x_m(t) - x_m(0) - \alpha t| \leq D_0$, one has

$$|\alpha_m - \alpha| \leq \frac{|x(t) - x_m(t) - x(0) + x_m(0)|}{t} + \frac{2D_0}{t} .$$

Given $\epsilon > 0$ choose $t$ so large that $2D_0/t < \epsilon/2$ and then $m$ so that

$$\frac{|x(t) - x_m(t) - x(0) + x_m(0)|}{t} \leq \epsilon/2$$

in $C(K)$, where $K = [-T, T]$ is a compact interval which contains 0 and $t$. Therefore $|\alpha - \alpha_m| < \epsilon$. □

**Corollary 2.4.9.** $\bigcup_{|\alpha| \leq A} \mathcal{M}_\alpha / \mathbb{Z}$ *is compact in the* $C^1(\mathbb{R})$ *topology.*

*Proof.* The fact that $\bigcup_{|\alpha| \leq A} \mathcal{M}/\mathbb{Z}$ is relatively compact in $C^1(\mathbb{R})$ follows from the theorem of Arzela–Ascoli and Theorem 2.4.7. To show compactness we need to show the closedness in $C^1(\mathbb{R})$. Let $\gamma_m$ be a sequence in $\bigcup_{|\alpha| \leq M} \mathcal{M}_\alpha / \mathbb{Z}$ with $\gamma_m \to \gamma \in C^1(\mathbb{R})$ in the $C^1$ topology. We claim that $\gamma \in \bigcup_{|\alpha| \leq M} \mathcal{M}_\alpha / \mathbb{Z}$.

1) $\gamma : t \mapsto x(t) \in \mathcal{M}$. Otherwise, there would exist a function $\phi \in C^1_{comp}(\mathbb{R})$ with support in $[-T, T]$ satisfying

$$\int_{-T}^{T} F(t, x + \phi, \dot{x} + \dot{\phi}) \, dt < \int_{-T}^{T} F(t, x, \dot{x}) \, dt \ .$$

Because of the uniform convergence $x_m \to x$, $\dot{x}_m \to \dot{x}$ on $[-T, T]$ we know that for sufficiently large $m$ also

$$\int_{-T}^{T} F(t, x_m + \phi, \dot{x}_m + \dot{\phi}) \, dt < \int_{-T}^{T} F(t, x_m, \dot{x}_m) \, dt$$

holds. This is a contradiction.

2) The fact that $\gamma \in \bigcup_{|\alpha| \leq M} \mathcal{M}_\alpha / \mathbb{Z}$ follows from the continuity in $\alpha$ if the $C^1$ topology is chosen on $\mathcal{M}$. (We would even have continuity in the weaker $C^0$ topology by Lemma 2.4.8.) $\qquad\square$

We know from Lemma 2.4.6 already that $\mathcal{M} \supset \mathcal{M}(p/q) \neq \emptyset$ and that $\mathcal{M}_\alpha \neq \emptyset$ for rational $\alpha$. With Corollary 2.4.9 the existence of minimals with irrational rotation number is established:

**Theorem 2.4.10.** $\mathcal{M}_\alpha \neq \emptyset$ *for every* $\alpha \in \mathbb{R}$.

*Proof.* Given $\alpha \in \mathbb{R}$, there exists a sequence $\{\alpha_m\} \subset \mathbb{Q}$ with $\alpha_m \to \alpha$.

For every $m$ choose an element $\gamma_m \in \mathcal{M}_{\alpha_m} \subset \bigcup_{|\beta| \subset A} \mathcal{M}_\beta / \mathbb{Z}$ with $\alpha < A$.
By the compactness obtained in Corollary 2.4.9 there is a subsequence of $\gamma_m \in \mathcal{M}_{\alpha_m}$ which converges to an element $\gamma \in \mathcal{M}_\alpha$. $\qquad\square$

## 2.5   $\mathcal{M}_\alpha$ for irrational $\alpha$, Mather sets

If $\alpha$ is irrational and $\gamma \in \mathcal{M}_\alpha$, $\gamma : t \mapsto x(t)$, we have a **total order** on the fundamental group $\mathbb{Z}^2$ of $\mathbb{T}^2$ defined by

$$(j, k) < (j', k') \Leftrightarrow x(j) - k < x(j') - k' \ .$$

It has also the property

$$(j,k) = (j',k') \Leftrightarrow x(j) - k = x(j') - k' .$$

Proof. If $x(j) - k = x(j') - k'$, then $x(t+q) - p = x(t)$ with $q = j' - j$ and $p = k' - k$, which means $q = p = 0$ or $\alpha = p/q$. Because $\alpha$ is irrational, $(j,k) = (j',k')$ follows. This order is the same as the order defined by $F = p^2/2$:

$$(j,k) < (j',k') \Leftrightarrow \alpha j - k < \alpha j' - k' .$$

Let $S_t := \{\alpha(j+t) - k \mid (j,k) \in \mathbb{Z}^2\}$ and $S = \{(t,\theta) \mid \theta = \alpha(j+t) - k \in S_t, t \in \mathbb{R}, (j,k) \in \mathbb{Z}^2\}$. We define the map

$$u : S \to \mathbb{R}, (t, \theta = \alpha(j+t) - k) \mapsto x(j+t) - k .$$

---

**Theorem 2.5.1.** a) $u$ is strictly monotone in $\theta$:

$$\alpha(j+t) - k < \alpha(j'+t) - k' \Leftrightarrow x(j+t) - k < x(j'+t) - k' .$$

b) $u(t+1, \theta) = u(t, \theta)$.
c) $u(t, \theta+1) = u(t, \theta) + 1$.

---

Proof. a) $\alpha(j+t) - k < \alpha(j'+t) - k' \Leftrightarrow x(j+t) - k < x(j'+t) - k'$ is with $q = j' - j$ and $p = k' - k$ equivalent to

$$0 < \alpha q - p \Leftrightarrow x(t) < x(t+q) - p .$$

We can assume $q > 0$ because otherwise $(j,k)$ could be replaced with $(j',k')$ and "<" with ">".

i) From $x(t) < x(t+q) - p$ we obtain by induction for all $n \in \mathbb{N}$:

$$x(t) < x(t+nq) - np ,$$

or after division by $nq$,

$$\frac{x(t)}{nq} < \frac{x(t+nq)}{nq} - \frac{p}{q} .$$

The limit $n \to \infty$ gives

$$0 \leq \alpha - \frac{p}{q} .$$

Because $\alpha$ is irrational, we have $\alpha q > p$.

ii) For the reversed implication we argue indirectly: from $x(t) \geq x(t+q) - p$ we get, proceeding as in i), also $\alpha < p/q$.

b) For $\theta = a(j + t) - k$ we have

$$u(t+1,\theta) = u(t+1,\alpha(j+t)-k) = u(t+1,\alpha(j-1+t+1)-k) = x(t+j)-k = u(t,\theta) \ .$$

c) $u(t,\theta + 1) = u(t,\alpha(j + t) - k + 1) = x(t + j) - k + 1 = u(t,\theta) + 1.$    □

For $t = 0$ we obtain

$$u(0,\theta + \alpha) = x(j + 1) - k = f(x(j) - k) = f(u(0,\theta)) \ ,$$

Therefore, with $u_0(\theta) = u(0,\theta)$,

$$u_0(\theta + \alpha) = f \circ u_0(\theta) \ .$$

The map $f$ is therefore conjugated to a rotation by the angle $\alpha$. However $u$ is defined on $S$, a dense subset of $\mathbb{R}$. If $u$ could be extended continuously to $\mathbb{R}$, then, by the monotonicity proven in Theorem 2.5.1, it would be a homeomorphism and $f$ would be conjugated to a rotation.

By closure we define two functions $u^+$ and $u^-$:

$$u^+(t,\theta) \quad = \quad \lim_{\theta_n \to \theta, \theta_n > \theta} u(t,\theta_n) \ ,$$
$$u^-(t,\theta) \quad = \quad \lim_{\theta_n \to \theta, \theta_n < \theta} u(t,\theta_n) \ .$$

There are two cases:

**case A):** $u^+ = u^- = u$ (which means $u$ is continuous).
**case B):** $u^+ \neq u^-$.

In the first case, $u = u(t,\theta)$ is continuous and strictly monotone in $\theta$: indeed if $\theta < \theta'$, there exist $(j,k)$ and $(j',k')$ with

$$\theta < (t + j)\alpha - k < (t + j')\alpha - k' < \theta'$$

and therefore also with Theorem 2.5.1 a),

$$u(t,\theta) \leq u(t,(t + j)\alpha - k) < u(t,(t + j')\alpha - k') \leq u(t,\theta')$$

and we have the strict monotonicity. This means that the map

$$h : (t,\theta) \to (t, u(t,\theta))$$

is a homeomorphism on the plane $\mathbb{R}^2$. It can be interpreted as a homeomorphism on the torus because it commutes with every map

$$(t,\theta) \mapsto (t + j, \theta + k) \ .$$

For every $\beta \in \mathbb{R}$ we have $\gamma_\beta \in \mathcal{M}_\alpha$, where

$$\gamma_\beta : t \mapsto x(t, \beta) = u(t, \alpha t + \beta)$$

satisfies $x(t, \beta) < x(t, \beta')$ for $\beta < \beta'$. We have therefore a one-parameter family of extremals.

**Question.** Is this an extremal field? Formal differentiation gives

$$\frac{d}{dt} x(t, \beta) = (\partial_t + \alpha \partial_\theta) u(t, \theta) = (\partial_t + \alpha \partial_\theta) u h^{-1}(t, x) .$$

In order to have an extremal field, we have to establish that

$$\psi(t, x) = (\partial_t + \alpha \partial_\theta) u h^{-1}(t, x) = \dot{x}(t, \beta)$$

is continuously differentiable. This is not the case in general. Nevertheless, we can say:

---

**Theorem 2.5.2.** *If $\alpha$ is irrational, $|\alpha| \leq A$ and $\gamma : t \mapsto x(t) \in \mathcal{M}_\alpha$ and if we are in the case A), then $\psi = (\partial_t + \alpha \partial_\theta) u h^{-1} \in \mathrm{Lip}(\mathbb{T}^2)$.*

---

*Proof.* (The proof requires Theorem 2.5.3 below). First of all, $\psi$ is defined on the torus because

$$\psi(t + 1, x) = \psi(t, x) = \psi(t, x + 1) .$$

To verify the Lipschitz continuity we have to find a constant $L$ such that

$$|\psi(t', x') - \psi(t'', x'')| \leq L(|t' - t''| + |x' - x''|) .$$

For $x' = x(t', \beta')$ and $x'' = x(t'', \beta'')$ we introduce a third point $y = x(t', \beta'')$.

$$
\begin{aligned}
|\psi(t', y) - \psi(t'', x'')| &= |\dot{x}(t', \beta'') - \dot{x}(t'', \beta'')| \\
&\leq |t' - t''| C_2(A) , \\
|\psi(t', x') - \psi(t', y)| &= |\dot{x}(t', \beta') - \dot{x}(t', \beta'')| \\
&\leq M(A)|x' - y| \quad \text{(Theorem 2.5.3)} \\
&\leq M(A)|x' - x''| , \\
|\psi(t', x') - \psi(t'', x'')| &< |\psi(t', y) - \psi(t'', x'')| + |\psi(t', x') - \psi(t', y)| \\
&\leq L(A)(|t' - t''| + |x' - x''|) ,
\end{aligned}
$$

with $L(A) = \max\{C_2(A), M(A)\}$, where $M(A)$ is defined below. In the first step of the second equation we have used the following Theorem 2.5.3. $\square$

**Theorem 2.5.3.** *Let* $\gamma, \eta \in \mathcal{M}_\alpha$, $\gamma : t \mapsto x(t), \eta : t \mapsto y(t)$, $x(t) > y(t)$, $|\alpha| \leq A$ *with* $A > 1$. *There is a constant* $M = M(A)$ *such that* $|\dot{x} - \dot{y}| \leq M|x - y|$ *for all* $t \in \mathbb{R}$.

*Proof.* For $x, y \in \mathcal{M}[-T, T]$, Lemma 2.3.4 assures that

$$|\dot{x}|, |\dot{y}| < C_1(A) = c_1 A^2 T^{-1} \,.$$

Let $\xi(t) = x(t) - y(t) > 0$ in $[-T, T]$. It is enough to show

$$|\dot{\xi}(0)| < M|\xi(0)|$$

because of the invariance of the problem with respect to time translation. Subtraction of the two Euler equations

$$\frac{d}{dt}F_p(t, x, \dot{x}) - F_x(t, x, \dot{x}) = 0 \,,$$
$$\frac{d}{dt}F_p(t, y, \dot{y}) - F_x(t, y, \dot{y}) = 0 \,,$$

gives

$$\frac{d}{dt}(A_0\dot{\xi} + B\xi) - (B\dot{\xi} + C\xi) = 0$$

with

$$A_0 = \int_0^1 F_{pp}(t, x + \lambda(y - x), \dot{x} + \lambda(\dot{y} - \dot{x}) \, d\lambda \,,$$

$$B = \int_0^1 F_{px}(t, x + \lambda(y - x), \dot{x} + \lambda(\dot{y} - \dot{x})) \, d\lambda \,,$$

$$C = \int_0^1 F_{xx}(t, x + \lambda(y - x), \dot{x} + \lambda(\dot{y} - \dot{x}) \, d\lambda \,.$$

By assumptions (i) and (ii) in Section 2.1, we conclude

$$\delta \leq A_0 \leq \delta^{-1} \,,$$
$$|B| \leq \lambda \,,$$
$$|C| \leq \lambda^2 \,,$$

with $\lambda = c_0 A^2 T^{-1}$, where $c_0$ is an $F$ dependent constant $\geq 1$ and $A \geq 1$ is a bound for $|\alpha|$ and $|x(T) - x(-T)|, |y(T) - y(-T)|$. With the following lemma, the proof of Theorem 2.5.3 is done.  □

**Lemma 2.5.4.** *Let $\xi = \xi(t)$ be in $[-T, T]$ a positive solution of the Jacobi equation $\frac{d}{dt}(A_0\dot{\xi} + B\xi) = B\dot{\xi} + C\xi$. Then,*

$$|\dot{\xi}(0)| \leq M\xi(0) ,$$

*where $M = 5c_0 A^2 T^{-1} \delta^{-2}$.*

*Proof.* Because $\xi > 0$ for $t \in [-T, T]$, we can form

$$\eta := A_0 \frac{\dot{\xi}}{\xi} + B .$$

For $t = -\tau$ we get

$$
\begin{aligned}
\frac{d}{d\tau}\eta &= -\dot{\eta} = -\frac{d}{dt}\left(\frac{A_0\dot{\xi} + B\xi}{\xi}\right) \\
&= \frac{\frac{d}{dt}(A_0\dot{\xi} + B\xi)}{\xi} + \frac{\dot{\xi}}{\xi^2}(A_0\dot{\xi} + B\xi) \\
&= -\frac{B\dot{\xi} + C\xi}{\xi} + A_0\left(\frac{\dot{\xi}}{\xi}\right)^2 + B\frac{\dot{\xi}}{\xi} \\
&= A_0^{-1}(\eta^2 - 2B\eta + B^2 - A_0 C)
\end{aligned}
$$

and so

$$\frac{d}{d\tau}\eta = A_0^{-1}(\eta - B)^2 - C .$$

This quadratic differential equation is called **Riccati equation**. We want to estimate $|\eta(0)|$. In our case we can assume $\eta(0) > 0$ because if we replace $(t, h)$ by $(-t, -h)$ and $B$ by $(-B)$, the Riccati equation stays invariant.

**Claim.** $|\eta(0)| \leq 4\lambda\delta^{-1}$.
If the claim were wrong, then $\eta(0) > 4\lambda\delta^{-1}$. For $t > 0$, as long the solution exists, the relation

$$\eta(\tau) \geq \eta(0) > 4\lambda\delta^{-1}$$

follows. Indeed, for $\eta > 4\lambda\delta^{-1}$,

$$|2B\eta| + |B^2 - A_0 C| < 2\lambda\eta + \lambda^2(1 + \delta^{-1}) < 2\lambda\eta + 2\lambda^2\delta^{-1} < \frac{\delta\eta^2}{2} + \frac{\delta\eta^2}{4} = \frac{3}{4}\eta^2\delta$$

so that from the Riccati equation, we get

$$\frac{d\eta}{d\tau} \geq \delta(\eta^2 - \frac{3}{4}\eta^2\delta) \geq \delta\eta^2/4 > 0 .$$

This inequality leads not only to the monotonicity property, but also the comparison function

$$\eta(\tau) \geq \frac{\eta(0)}{1 - \eta(0)\delta\tau/4}$$

which is infinite for $t = 4\delta^{-1}\eta(0)^{-1}$. Therefore,

$$T < 4\delta^{-1}\eta(0)^{-1}$$

or $\eta(0) < 4T^{-1}\delta^{-1} \leq 4A^2T^{-1}\delta^{-1} = 4\lambda\delta^{-1}$ which contradicts our assumption. The claim $|\eta(0)| \leq 4\lambda\delta^{-1}$ is now proven. Because

$$\frac{\dot{\xi}}{\xi} = A_0^{-1}(\eta - B)$$

one has

$$\frac{|\dot{\xi}(0)|}{\xi(0)} \leq \delta^{-1}(4\lambda\delta^{-1} + \lambda) \leq 5\lambda\delta^{-2} = 5c_0A^2T^{-1}\delta^{-2} . \qquad \square$$

**Definition.** A **global Lipschitz-extremal field** on the torus is given by a vector field $\dot{x} = \psi(t, x)$ with $\psi \in \mathrm{Lip}(\mathbb{T}^2)$, so that every solution $x(t)$ is extremal.

Theorem 2.5.2 says that a minimal with irrational rotation number in case A) can be embedded into a global Lipschitz-extremal field.

**Example. Free pendulum.**

$F = \frac{1}{2}\left(p^2 - \frac{1}{\pi}\cos(2\pi x)\right)$ has the Euler equations

$$\ddot{x} = \frac{1}{2}\sin(2\pi x)$$

with the energy integral

$$E = \frac{\dot{x}^2}{2} + \frac{1}{4\pi}\cos(2\pi x) \geq -\frac{1}{4\pi} .$$

Especially for $E = (4\pi)^{-1}$ we get

$$\dot{x}^2 = \frac{1}{2\pi}(1 - \cos(2\pi x)) = \frac{1}{\pi}\sin^2(\pi x)$$

or

$$\dot{x} = \pm\sin(\pi x)/\sqrt{\pi}$$

and in order to get the period 1, we take

$$|\dot{x}| = |\sin(\pi x)/\sqrt{\pi}| = \psi(t, x) .$$

$\psi$ is not $C^1$ but Lipschitz continuous with Lipschitz constant $\sqrt{\pi}$.

In the Hamiltonian formulation, things are similar. Because Lipschitz surfaces have tangent planes almost everywhere only, we make the following definition:

**Definition.** A Lipschitz surface $\Sigma$ is called **invariant under the flow of $H$**, if the vector field

$$X_H = \partial_t + H_y \partial_x - H_x \partial_y$$

is almost everywhere tangential to $\Sigma$.

---

**Theorem 2.5.5.** *If $\dot{x} = \psi(t, x)$ a Lipschitz extremal field is for $F$, then*

$$\Sigma = \{(t, x, y) \in \Omega \times \mathbb{R} \mid y = F_p(t, x, \psi(t, x))\}$$

*is Lipschitz and invariant under the flow of $H$. On the other hand, if $\Sigma$ is a surface which is invariant under the flow of $H$ given by*

$$\Sigma = \{(t, x, y) \in \Omega \times \mathbb{R} \mid y = h(t, x)\},$$

*with $h \in \mathrm{Lip}(\Omega)$, then the vector field $\dot{x} = \psi(t, x)$ defined by*

$$\psi = H_y(t, x, h(t, x))$$

*is a Lipschitz extremal field.*

---

The mathematical pendulum from the first section had invariant $C^1$ tori. However for the energy $E = (4\pi)^{-1}$, the extremal field is only Lipschitz continuous.

While for irrational $\alpha$ and in case A), the construction of Lipschitz extremal fields has been established, the question appears whether there might be different $\psi \in \mathcal{M}_\alpha$, which can not be embedded into this extremal field. The answer is negative:

---

**Theorem 2.5.6.** *If $\gamma, \eta \in \mathcal{M}_\alpha$, $\gamma : t \mapsto x(t)$, $\eta : t \mapsto y(t)$ are given and $\alpha$ is irrational and if we are in case A), then there exists $\beta \in \mathbb{R}$, such that $y = u(t, \alpha t + \beta)$ and $\eta$ is in case A) too.*

---

The proof of Theorem 2.5.6 will be given later.

**Remarks.**
1) Theorem 2.5.6 states that all elements of $\mathcal{M}_\alpha$ belong to the extremal field, which is generated by $\gamma$ and that the decision to belong to case A) or B) does not depend on the element $\gamma \in \mathcal{M}_\alpha$.
2) In case A) there is for every $\alpha$ exactly one $\gamma \in \mathcal{M}(\alpha)$, with $x(0) = \alpha$. This follows from the existence and uniqueness theorem for ordinary differential equations.
3) In case A) every $\gamma \in \mathcal{M}_\alpha$ is dense in $\mathbb{T}^2$, because the map is a homeomorphism in this case.

What happens in case B)? Can it occur at all?

**Example.** Consider $F = \frac{1}{2}p^2 + V(t,x)$. Assume that
the torus is parameterized by $|x| \leq 1/2$, $|t| \leq 1/2$.
Define $V$ as a $C^\infty(\mathbb{T}^2)$-function for $0 < \rho < r \leq 1/6$:

$$V(t,x) \geq M \geq 1 \quad , \quad x^2 + t^2 \leq \rho^2 ,$$
$$V(t,x) = v(t^2 + x^2) \geq 0 \quad , \quad \rho^2 \leq x^2 + t^2 \leq r^2 ,$$
$$V(t,x) = 0 \quad , \quad x^2 + t^2 \geq r^2 .$$

> **Claim.** For every $\alpha \in \mathbb{R}$ with $\rho^2 M > 6(|\alpha| + 1|)^4$,
> case B) happens for $\mathcal{M}_\alpha$.

*Proof.* Assume that there exists an $\alpha \in \mathbb{R}$ with

$$\rho^2 M > 6[|a| + 1]^4$$

and that we were in case A). According to the above remark 3), there would be a
minimal $\gamma \in \mathcal{M}, \gamma : t \mapsto x(t)$ with $x(0) = 0$.

We will show now that $\gamma$ can not be minimal in the class of curves, which start
at $A := (t_1, a = x(t_1)) = (-0.5, x(-0.5))$ and end at $B = (t_1, b = x(t_2)) = (0.5, x(0.5))$. This will lead to a contradiction.

Since by Theorem 2.4.4, $|x(t+j) - x(t) - j\alpha| \leq 1$ for every $j \in \mathbb{Z}$, the inequality

$$m := |x(\frac{1}{2}) - x(-\frac{1}{2})| \leq 1 + |\alpha| .$$

Let $t_1$ and $t_2$ be chosen in such a way that

$$t_1 \quad < \quad 0 \quad < \quad t_2 ,$$
$$t^2 + x(t)^2 \quad \leq \quad \rho^2, \quad t \in [t_1, t_2] .$$

This means that the diameter $2\rho$ of $B_\rho = \{(t,x) \mid t^2 + x^2 \leq \rho^2 \}$ is smaller than
or equal to the length of $\gamma$ between $x(t_1)$ and $x(t_2)$:

$$2\rho \leq \int_{t_1}^{t_2} \sqrt{1 + \dot{x}^2} \, dt \leq \sqrt{t_2 - t_1} [\int_{t_1}^{t_2} (1 + \dot{x}^2) \, dt]^{1/2}$$

and therefore with $\tau = \tau_2 - \tau_1$,

$$\int_{t_1}^{t_2} (1 + \dot{x}^2) \, dt \geq \frac{4\rho^2}{\tau} .$$

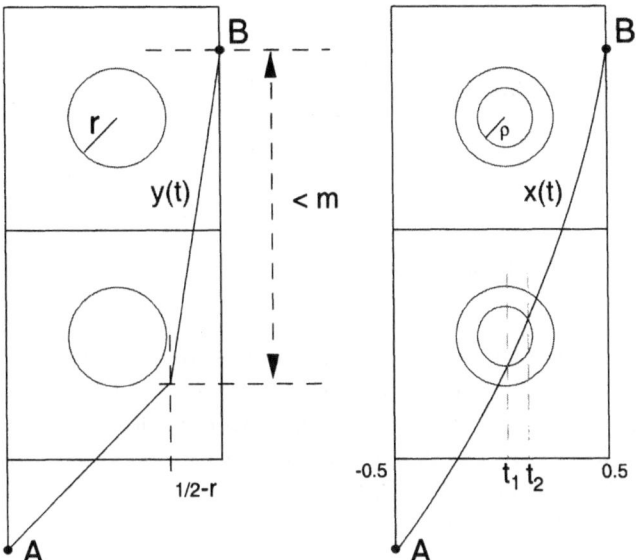

The action of $\gamma$ connecting $A$ with $B$ can now be estimated:

$$\int_{-1/2}^{1/2} F(t, x, \dot{x})\, dt \geq \int_{t_1}^{t_2} F(t, x, \dot{x})\, dt$$

$$\geq \int_{t_1}^{t_2} \frac{1}{2}(\dot{x}^2 + 1) + (M - \frac{1}{2})\, dt$$

$$\geq \frac{2\rho^2}{\tau} + (M - \frac{1}{2})\tau \ .$$

With the special choice $\tau = 2\rho[2M - 1]^{-1/2}$ we have

$$\int_{-1/2}^{1/2} F(t, x, \dot{x})\, dt \geq \frac{2\rho^2}{\tau} + (M - \frac{1}{2})\tau = 2\rho\sqrt{2M - \rho} \ .$$

We choose now a special path $\eta : t \mapsto y(t)$ inside the region where $V = 0$. This can be done with a broken line $t \mapsto y(t)$, where

$$\dot{y} \leq \frac{m}{(1/2 - r)} \leq 3m \ .$$

We have then

$$\int_{-1/2}^{1/2} F(t, y, \dot{y})\, dt \leq \int_{-1/2}^{1/2} \frac{\dot{y}^2}{2}\, dt \leq \frac{9}{2}m^2 \leq \frac{9}{2}(1 + |\alpha|)^2 \ .$$

By the minimality of $\gamma$ we have

$$2\rho\sqrt{2M - \rho} \leq \frac{9}{2}(1 + |\alpha|)^2$$

and so

$$4\rho^2(2M - \rho) \leq \frac{81}{4}(1 + |\alpha|)^4 \,,$$

$$4M\rho^2 \leq \frac{81}{4}(1 + |a|)^4 \,,$$

$$M\rho^2 \leq 6(1 + |a|)^4 \,,$$

which is a contradiction to the assumption.                    □

**Remarks**.
1) Because $V$ can be approximated arbitrarily well by real-analytic $V$, it is also clear that there exist real-analytic $V$ for which we are in case B).
2) Without giving a proof we note that in this example, for fixed $\rho, r, M$ and sufficiently large $\alpha$ we are always in case A). The reason is that for big $\alpha$, the summand $p^2/2$ has large weight with respect to $V(t, x)$. To do the $\alpha \to \infty$ limit in the given variational problem is equivalent to do the $\epsilon \to 0$ limit in the variational problem

$$F' = \frac{p^2}{2} + \epsilon V(t, x).$$

The latter is a problem in perturbation theory,  a topic in the so-called KAM theory in particular.

Let $\gamma \in M_\alpha$ and $\alpha$ irrational be given and assume that $M_\alpha$ is in case B). By definition we have $u^+ \neq u^-$, where $u^+$ and $u^-$ are the functions constructed from $\gamma$. For every $t$ the set $\{\theta \mid u^+(t, \theta) \neq u^-(t, \theta)\}$ is countable.

**Definition**.  Define the sets

$$\mathcal{M}_t^\pm := \{u^\pm(t, \theta) \mid \theta \in \mathbb{R}\}$$

and **the limit set** of the orbit $\gamma$

$$M(\gamma) = \{u^\pm(t, \theta) \mid t, \theta \in \mathbb{R}\} \,.$$

$\mathcal{M}_t := \mathcal{M}_t^+ \cap \mathcal{M}_t^-$ is the set of continuity of $u^+$ rsp. $u^-$. There are only countably many discontinuity points. An important result of this section is the following theorem:

---

**Theorem 2.5.7.** *Let $\alpha$ be irrational, $\gamma : t \mapsto x(t), \eta : t \mapsto y(t)$ both in $M(\alpha)$ with corresponding functions $u^\pm$ and $v^\pm$. Then there exists a constant $c \in \mathbb{R}$ such that $u^\pm(t, \theta) = v^\pm(t, \theta + c)$.*

---

*Proof.* 1) **It is enough to prove the claim for** $t = 0$.

We use the notation $u_0^\pm(\theta) = u^\pm(0, \theta)$. Assume, there exists $c$ with

$$u_0^\pm(\theta) = v_0^\pm(\theta + c), \forall \theta .$$

Then also

$$u_0^\pm(\theta + \alpha) = v_0^+(\theta + \alpha + c) .$$

Define for fixed $\theta$,

$$\tilde{\gamma} : t \quad \mapsto \quad \tilde{x}(t) := u^\pm(t, \alpha t + \theta) ,$$
$$\tilde{\eta} : t \quad \mapsto \quad \tilde{v}(t) := v^\pm(t, \alpha t + \theta + c) .$$

$\tilde{\gamma}$ and $\tilde{\eta}$ are in $\mathcal{M}_\alpha$. Because of the two intersections

$$\tilde{x}(1) \;\; = \;\; \tilde{y}(1) ,$$
$$\tilde{x}(0) \;\; = \;\; \tilde{y}(0) ,$$

the two curves $\tilde{\gamma}$ and $\tilde{\eta}$ are the same. Replacing $\alpha t + \theta$ with $\alpha t + \theta + c$ establishes the claim $u^\pm(t, \theta) = v^\pm(t, \theta + c)$ for all $t$.

2) **If for some** $\lambda \in \mathbb{R}$ **and some** $\theta \in \mathbb{R}$ **the conditions** $v_0^-(\theta + \lambda) - u_0^-(\theta) < 0$ **hold, then** $v_0(\theta + \lambda) - u_0(\theta) \leq 0$, $\forall \theta \in \mathbb{R}$.

Otherwise, $v_0^-(\theta + \lambda) - u_0^-(\theta)$ changes sign. By semicontinuity, there would exist intervals $I^+$ and $I^-$ of positive length, for which

$$v_0^-(\theta + \lambda) - u_0^-(\theta) > 0, \text{ in } I^+ ,$$
$$v_0^-(\theta + \lambda) - u_0^-(\theta) < 0, \text{ in } I^- .$$

We put

$$\tilde{x}(t) \;\; = \;\; u_0^-(t, \alpha t) ,$$
$$\tilde{y}(t) \;\; = \;\; v_0^-(t, \alpha t + \lambda) .$$

Then

$$\tilde{y}(j) - \tilde{x}(j) = \tilde{y}(j) - k - (\tilde{x}(j) - k) = v_0^-(\lambda + \alpha j - k) - u_0^-(\alpha j - k)$$

and this is positive, if $\alpha j - k \in I^+$ and negative if $\alpha j - k \in I^-$. Because $\{\alpha j - k, j, k \in \mathbb{Z} \}$ is dense in $\mathbb{R}$ there would be infinitely many intersections of $\tilde{x}$ and $\tilde{y}$. This is a contradiction.

3) $c := \sup\{\lambda \mid v_0^-(\theta + \lambda) - u_0^-(\theta) \leq 0, \forall \theta \}$ **is finite and the supremum is attained.**

There exists a constant $M$, so that for all $\theta \in \mathbb{R}$,

$$|v_0^-(\theta + \lambda) - (\theta + \lambda)| \le M \,,$$
$$|u_0^-(\theta) - \theta| \le M \,.$$

Because of Theorem 2.5.1, both functions on the left-hand side are periodic. Therefore

$$|v_0^-(\theta + \lambda) - u_0^-(\theta) - \lambda| \le 2M$$

or

$$v_0^-(\theta + \lambda) - u_0^-(\theta) \ge \lambda - 2M \,.$$

Because the left-hand side is $\le 0$, we have

$$\lambda \le 2M$$

and $c$ is finite. If a sequence $\lambda_n$ converges from below to $c$ and for all $n$

$$v_0^-(\theta + \lambda_n) - u_0^-(\theta) \le 0, \ \forall \theta \,,$$

then also

$$v_0^-(\theta + c) - u_0^-(\theta) \le 0, \ \forall \theta$$

because of the left semi continuity of $v_0^-$.

4) $v_0^-(\theta + c) - u_0^-(\theta) = 0$, **if $\theta + c$ is a point of continuity of $v_0^-$.**

Otherwise there would exist $\theta^*$ with

$$v_0^-(\theta^* + c) - u_0^-(\theta^*) < 0 \,,$$

where $\theta^* + c$ is a point of continuity. This implies that there exists $\lambda > c$ with

$$v_0^-(\theta^* + \lambda) - u_0^-(\theta^*) < 0 \,.$$

With claim 2) we conclude that

$$v_0^-(\theta + \lambda) - u_0^-(\theta) \le 0, \ \forall \theta \,.$$

This contradicts the minimality of $\gamma$.

5) $v_0^\pm(\theta + c) = u_0^\pm(\theta), \forall \theta.$

Having only countably many points of discontinuity the functions $v_0^+$ and $u_0^-$ are uniquely determined by the values at the places of continuity:

$$v_0^-(\theta + c) = u_0^-(\theta), \forall \theta \,.$$

Because $v_0^+ = v_0^-$ and $u_0^+ = u_0^-$ at the places of continuity,

$$v_0^+(\theta + c) = u_0^+(\theta), \forall \theta \,. \qquad \square$$

In the next theorem the **gap size**

$$\xi(t) = x^+(t) - x^-(t) = u^+(t, \alpha t + \beta) - u^-(t, \alpha t + \beta)$$

is estimated:

---

**Theorem 2.5.8.** *Let $|\alpha| \le A$ and let $M(A)$ be the constant of Theorem 2.5.3. There exists a constant $C = C(A) = \log(M(A))$, with*

$$\exp(-C|t - s|) \le \xi(t)/\xi(s) \le \exp(C|t - s|) \, .$$

---

*Proof.* According to Theorem 2.5.3 the relation

$$|\dot\xi(t)| \le M\xi(t)$$

holds and therefore

$$
\begin{aligned}
|\dot\xi|/\xi &\le M \, , \\
|\frac{d}{dt} \log \xi| &\le M \, , \\
|\log \xi(t) - \log \xi(s)| &\le M|t - s| \, . \qquad \square
\end{aligned}
$$

---

**Theorem 2.5.9.** *For irrational $\alpha$, the set $\mathcal{M}_\alpha$ is totally ordered: $\forall \gamma, \eta \in \mathcal{M}_\alpha$ we have $\gamma < \eta$ or $\gamma = \eta$ or $\gamma > \eta$.*

---

**Remarks.**
1) Theorem 2.5.9 says that two minimals with the same rotation number do not intersect.

2) As we will see in the next section, this statement is wrong for $\alpha \in \mathbb{Q}$, where pairs of intersecting orbits, so called **homoclinic orbits** can exist.

3) Still another formulation of Theorem 2.5.9 would be: the projection

$$p : \mathcal{M}_\alpha \to \mathbb{R}, x \mapsto x(0)$$

is injective. This means that for every $a \in \mathbb{R}$ there exists at most one $x \in \mathcal{M}$ with $x(0) = a$. Only in case A), the projection $p$ is also surjective.

4) Theorem 2.5.9 implies Theorem 2.5.5.

*Proof of Theorem 2.5.9.* We use that for $x \in \mathcal{M}_\alpha$ the set of orbits

$$\{\gamma_{jk} : x(t + j) - k \}$$

and therefore also their closure $\mathcal{M}(x)$ is totally ordered.
Because by definition $u^-(t, \alpha t + \beta) \in \mathcal{M}_\alpha(x)$, the claim follows for

$$y(t) = u^\pm(t, \alpha t + \beta) \,.$$

It remains the case, where $y$ is itself in a gap of the Mather set of $x$:

$$u^-(0, \beta) < y(0) < u^+(0, \beta) \,.$$

Because by Theorem 2.5.7 the functions $u^\pm$ are also generated by $y$, we know that
for all $t$,

$$u^-(t, \alpha t + \beta) < y(t) < u^-(t, \alpha t + \beta) \,.$$

We need to show the claim only if both $x$ and $y$ are in the same gap of the Mather
set. Let therefore

$$u^-(0, \beta) < x(0) \leq y(0) < u^+(0, \beta) \,.$$

We claim that the gap size

$$\xi(t) = u^+(t, \alpha t + \beta) - u^-(t, \alpha t + \beta) > 0$$

converges to 0 for $t \to \infty$. This would mean that $x$ and $y$ are asymptotic. With
Theorem 2.5.3 also $|\dot{x} - \dot{y}| \to 0$ and we would be finished by applying Theorem 2.3.1
c). The area of the gap

$$\int_\mathbb{R} \xi(t) \, dt \leq \mu(\mathbb{T}^2)$$

is finite because $\mu(\mathbb{T}^2)$ is the area of the torus. From Theorem 2.5.8 we know that
for $t \in [n, n+1)$,

$$M^{-1} \leq \xi(t)/\xi(n) \leq M \,.$$

Because

$$\sum_{n \in \mathbb{N}} \xi(n) \leq M \int_\mathbb{R} \xi(t) \, dt < \infty \,,$$

we have $\lim_{n \to \infty} \xi(n) = \lim_{t \to \infty} \xi(t) = 0.$                                       □

The question is left open whether there are minimal orbits in the gaps of the
Mather sets. Instead we characterize the orbits of the form

$$x(t) = u^\pm(t, \alpha t + \beta) \,.$$

Let

$$\mathcal{U}_\alpha = \{ x \in \mathcal{M}_\alpha \mid \exists \beta \ x(t) = u^\pm(\alpha t + \beta) \} \,.$$

**Definition.** An extremal solution $x(t)$ is called **recurrent**, if there exist sequences
$j_n$ and $k_n$ with $j_n \to \infty$, so that $x(t + j_n) - k_n - x(t) \to 0$ for $n \to \infty$. Denote
the set of recurrent minimals with $\mathcal{M}^{rec}$ and define $\mathcal{M}^{rec}_\alpha := \mathcal{M}^{rec} \cap \mathcal{M}_\alpha$.

---

**Theorem 2.5.10.** *For irrational $\alpha$ we have $\mathcal{U}_\alpha = \mathcal{M}^{rec}_\alpha$.*

---

*Proof.* (i) $\mathcal{U}_\alpha \subset \mathcal{M}_\alpha^{rec}$.
If $x \in \mathcal{U}_\alpha, x = u^+(t, \alpha t + \beta)$, then

$$x(t + j_n) - k = u^+(t, \alpha t + \beta + \alpha j_n - k_n)$$

and it is enough to find sequences $j_n, k_n$ with $\alpha j_n - k_n \to 0$. Therefore, $x$ is recurrent. In the same way the claim is verified for $x(t) = u^-(t, \alpha t + \beta)$.

(ii) $\mathcal{M}_\alpha^{rec} \subset \mathcal{U}_\alpha$.
We assume, $x \in \mathcal{M}_\alpha \setminus \mathcal{U}_\alpha$. This means that $x$ is recurrent and it is in a gap

$$u^-(0, \beta) < x(0) < u^+(0, \beta),$$
$$x(j_n) - k_n \to x(0), j_n \to \infty.$$

By the construction of $u^\pm(0, \beta)$ we have

$$x(j_n) - k_n \to u^\pm(0, \beta)$$

and therefore $x(0) = u^\pm(0, \beta)$. This is a contradiction. $\square$

**Definition.** Define $\mathcal{M}_\alpha^{rec}(\gamma) := \mathcal{M}_\alpha(\gamma) \cap \mathcal{M}_\alpha^{rec}$.

---

**Theorem 2.5.11.** *If $\alpha$ is irrational, then for all $\gamma_1$ and $\gamma_2 \in \mathcal{M}_\alpha$,*

$$\mathcal{M}_\alpha^{rec}(\gamma_1) = \mathcal{M}_\alpha^{rec}(\gamma_2) = \mathcal{M}_\alpha^{rec}.$$

---

*Proof.* According to Theorem 2.5.10 we have $\mathcal{M}_\alpha^{rec} = \mathcal{U}_\alpha$ and by construction we get $\mathcal{M}_\alpha^{rec}(\gamma) = \mathcal{U}_\alpha$. Theorem 2.5.7 assures that $\mathcal{U}_\alpha$ is independent of $\gamma$. $\square$

For every $(j, k) \in \mathbb{Z}^2$, let

$$T_{j,k} : \mathcal{M} \to \mathcal{M}, \ x(t) \mapsto x(t + j) - k.$$

$\mathcal{M}_\alpha$ and therefore also $\mathcal{M}_\alpha^{rec}$ is invariant under $T_{j,k}$. Which are the smallest, nonempty and closed subsets of $\mathcal{M}_\alpha$ which are $T_{j,k}$-**invariant**, that is invariant under all $T_{j,k}$?

---

**Theorem 2.5.12.** *In $\mathcal{M}_\alpha$, there is exactly one smallest non-empty $T_{j,k}$-invariant closed subset: it is $\mathcal{M}_\alpha^{rec}$.*

---

*Proof.* $\mathcal{M}_\alpha^{rec}$ is $T_{j,k}$-invariant, closed and not empty. Let $\mathcal{M}_\alpha^* \subset \mathcal{M}_\alpha$ have the same properties and let $x^* \in \mathcal{M}_\alpha^*$. Because of the closedness and invariance of $\mathcal{M}_\alpha^{rec}(x^*) \subset \mathcal{M}_\alpha^*$ and because of Theorem 2.5.11, also $\mathcal{M}_\alpha^{rec} \subset \mathcal{M}_\alpha^*$. $\square$

We know $\mathcal{M}_\alpha$ for irrational $\alpha$ by approximation by periodic minimals. We can now show that every recurrent minimal can be approximated by periodic minimals.

---

**Theorem 2.5.13.** *Every $x \in \mathcal{M}_\alpha^{rec}$ can be approximated by periodic orbits in $\mathcal{M}$.*

---

*Proof.* The set $\mathcal{M}^*$ of orbits which can be approximated by periodic minimals is $T_{j,k}$-invariant, closed and not empty. Because of Theorem 2.5.12 we have $\mathcal{M}_\alpha^{rec} \subset \mathcal{M}^*$. $\qquad \square$

**Definition.** In case B) one calls the elements in $\mathcal{M}_\alpha^{rec}$ **Mather sets**.

Mather sets are **perfect sets**. They are closed, nowhere dense sets for which every point is an accumulation point. A perfect set is also called a **Cantor set**.

Let us summarize the central statement of this section:

---

**Theorem 2.5.14.** *For irrational $\alpha$, the following holds:*
**case A):** *All minimal $x \in \mathcal{M}_\alpha$ are dense on the torus. This means that for all $(t,a) \in \mathbb{R}^2$, there exists a sequence $(j_n, k_n) \in \mathbb{Z}^2$ with $x(t + j_n) - k_n \to a$.*
**case B):** *no minimal $\gamma \in \mathcal{M}_\alpha$ is dense on the torus. In other words if $u^-(0, \beta) < a < u^+(0, \beta)$, then $(0, a)$ is never an accumulation point of $x$.*

---

We know that both cases A) and B) can occur. It is a delicate question to decide in which of the cases we are. The answer can depend on how well $\alpha$ can be approximated by rational numbers.

**Appendix: Denjoy theory.**
The theory developed so far is related with **Denjoy theory** from the first third of the twentieth century. We will state the main results of this theory without proofs.

Let $f$ be an **orientation preserving homeomorphism** on the circle $\mathbb{T}$. The following **Lemma of Poincaré** should be compared with Theorem 2.4.1.

---

**Lemma 2.5.15.** *The rotation number $\alpha(f) = \lim_{n \to \infty} f^n(t)/n$ exists and is independent of $t$.*

---

Let $S_t = \{\alpha(j+t) - k \mid (j,k) \in \mathbb{Z}^2 \}$ and $S = \{(t, \theta) \mid \theta = \alpha(j+t) - k \in S_t, t \in \mathbb{R} \}$ and define
$$u : S \to \mathbb{R} \, , (t, \theta = \alpha(j + t) - k) \to f^j(t) - k \,.$$

The next theorem should be compared with Theorem 2.5.1.

**Theorem 2.5.16.** *a) $u$ is strictly monotone in $\theta$. This means*

$$\alpha(j+t) - k < a(j'+t) - k' \Leftrightarrow f^j(t) - k < f^{j'}(t) - k' \ .$$

*b) $u(t+1, \theta) = u(t, \theta)$.*
*c) $u(t, \theta+1) = u(t, \theta) + 1$.*

Again we define by closure two functions $u^+$ and $u^-$ :

$$u^+(t, \theta) \quad = \quad \lim_{\theta < \theta_n \to \theta} u(t, \theta_n) \ ,$$

$$u^-(t, \theta) \quad = \quad \lim_{\theta > \theta_n \to \theta} u(t, \theta_n) \ .$$

There are two cases:

**case A):** $u^+ = u^- = u$   ($u$ is continuous) ,
**case B):** $u^+ \neq u^-$ .

The set

$$\mathcal{L}^\pm(t) := \{\omega \in \mathbb{T} \mid \exists j_n \to \pm\infty, f^{j_n}(t) \to \omega \}$$

is closed and $f$-invariant. The following theorem of Denjoy (1932) should be compared to Theorems 2.5.10, 2.5.11 and 2.5.12.

**Theorem 2.5.17.** *If $\alpha$ is irrational, then $\mathcal{L} = \mathcal{L}^+(t) = \mathcal{L}^-(t)$ is independent of $t$ and the smallest non-empty $f$-invariant, closed subset of $T$. In the case A) we have $\mathcal{L} = \mathbb{T}$, in the case B) the set $\mathcal{L}$ is a perfect set. If $f'$ is of bounded variation, we are in case A). For $f \in C^1$ it provides examples, where we are in case B).*

In case B) one calls the set $\mathcal{L}$ a **Denjoy-minimal set**. We see now the relations:

> *The intersection of a Mather set with the line $t = t_0$ is a Denjoy-minimal set for the continuation $f$ of the map $\tilde{f} : x(j) - k \to x(j+1) - k$ onto the circle.*

## 2.6   $\mathcal{M}_\alpha$ for rational $\alpha$

Let $\alpha = p/q$ with $q \neq 0$. We have seen in Lemma 2.4.6 of Section 2.4 and Theorem 2.2.2 that

$$\mathcal{M}_{p/q} \supset \mathcal{M}(q, p) \neq \emptyset \ .$$

$\mathcal{M}(p/q) = \mathcal{M}(q, p)$ is the set of minimal periodic orbits of type $(q, p)$.

**Question.** Is $\mathcal{M}_{p/q} = \mathcal{M}(p/q)$? No! Indeed there are pairs of orbits in $\mathcal{M}_{p/q}$, which intersect once and which can therefore not be contained in the totally ordered set $\mathcal{M}(p/q)$.

**Example.**

1) $F = p^2/2$, $\ddot{x} = 0$, $x(t) = \alpha t + \beta$. In this case we have $\mathcal{M}_{p/q} = \mathcal{M}(p/q)$.

2) $F = p^2/2 + \cos(2\pi x)$ , $E = \dot{x}^2/2 + \cos(2\pi x)$ is constant. Take $\alpha = 0$. We have $\mathcal{M}_0 \neq M(0)$, because $\mathcal{M}(0)$ is not totally ordered and $\mathcal{M}$ is totally ordered by Theorem 2.5.9. Note that $\mathcal{M}(0)$ is **not** well ordered because there are seperatrices with energy $E = (4\pi)^{-1}$ defined by

$$\dot{x} = \pm|\sin(\pi x)|/\sqrt{\pi} \ .$$

They both have zero rotation number and they intersect.

**Definition.** Two periodic orbits $x_1 < x_2 \in \mathcal{M}(p/q)$ are called **neighboring** if there exists no $x \in \mathcal{M}(p/q)$ with $x_1 < x < x_2$.

Note that $\mathcal{M}(p/q)$ is well ordered, justifying the above definition.

---

**Theorem 2.6.1.** *Let* $\gamma \in \mathcal{M}_{p/q}$. *There are three possibilities:*
*a)* $\gamma \in \mathcal{M}(p/q)$, *therefore* $x(t+q) - p = x(t)$.
*b$^+$)* *There are two neighboring periodic minimals* $\gamma_1 > \gamma_2$, $\gamma_i \in \mathcal{M}(p/q) : \gamma_i : t \mapsto x_i(t)$, *so that*
$x_1(t) - x(t) \to 0$ *for* $t \to \infty$ *and*
$x_2(t) - x(t) \to 0$ *for* $t \to -\infty$.
*b$^-$)* *There are two neighboring periodic minimals* $\gamma_1 > \gamma_2$, $\gamma_i \in \mathcal{M}(p/q) : \gamma_i : t \mapsto x_i(t)$, *so that*
$x_2(t) - x(t) \to 0$ *for* $t \to \infty$ *and*
$x_1(t) - x(t) \to 0$ *for* $t \to -\infty$.

---

*Proof.* Let $\gamma \in \mathcal{M}_{p/q}, \gamma : t \mapsto x(t)$ but in $\gamma \notin \mathcal{M}(p/q)$. Therefore, for all $t$

$$(i) \ x(t+q) - p \ > \ x(t) \ \text{or}$$
$$(ii) \ x(t+q) - p \ < \ x(t) \ .$$

We will show that (i) implies case b$^+$). By (i) the sequence

$$y_j(t) = x(t + jq) - pj$$

is monotonically increasing for $j \to \infty$ and bounded because of the estimate

$$|y_j(t) - y_j(0)| \leq C_0 \ .$$

Therefore $y_j$ converge to a function $x_2(t)$ which is again in $\mathcal{M}_{p/q}$. It is even periodic and of type $(q, p)$ and therefore an element in $\mathcal{M}(p/q)$. In the same way, $y_j$ converges for $j \to \infty$ to a function $x \in \mathcal{M}(p/q)$. We still have to show that $x_1$ and $x_2$ are neighboring. Let $\gamma^* : x^* \in \mathcal{M}(p/q)$ with $x_1 < x^* < x_2$ and call

$A = (t_0, x(t_0)) = (t_0, x^*(t_0))$ the now mandatory intersection of $x^*$ with $x$. We define also the points $B = (t_0+q, x^*(t_0+q))$, $P = (T-q, x(T-q))$ and $Q = (T, z(T))$, where $z : t \mapsto x(t-q)$ and $T > t_0 + q$. The new curves

$$\tilde{x}_1^*(t) := \begin{cases} x^*(t), & t \in [t_0, t_0 + q], \\ z(t) & t \in [t_0 + q, T], \end{cases}$$

$$\tilde{x}_2^*(t) := \begin{cases} x(t), & t \in [t_0, T - q], \\ w(t) & t \in [T - q, T], \end{cases}$$

with $w(t) = (T - t)x(t) - (T - q - t)z(t)$ are concurrent in the class of curves between $A$ and $Q$. We have

$$\int_{t_0}^{T-q} F(t, x, \dot{x}) \, dt = \int_{t_0+q}^{T} F(t, z, \dot{z}) \, dt$$

and

$$\int_{T-q}^{T} F(t, w, \dot{w}) \, dt \quad =_{T \to \infty} \quad \int_{T-q}^{T} F(t, \tilde{x}_2, \dot{\tilde{x}}_2) \, dt$$

$$= \quad \int_{t_0}^{t_0+q} F(t, \tilde{x}_2, \dot{\tilde{x}}_2) \, dt$$

$$= \quad \int_{t_0}^{t_0+q} F(t, x^*, \dot{x}^*) \, dt \, .$$

(The first equality in the last equation holds asymptotically for $T \to \infty$. The second equality is a consequence of the periodicity of $x_2$. The last equation follows from the minimality of $x_2$ and $x^*$ in $\mathcal{M}(p/q)$.)

Therefore, for $T \to \infty$, the actions of $\tilde{x}_1$ and $\tilde{x}_2$ between $A$ and $Q$ are approximatively equal. However, the action of the path $t \mapsto \tilde{x}_1(t)$ can be decreased at $B$ by a fixed and $T$-independent amount because $y$ has a corner there.

Therefore $x^*$ can not be minimal between $A$ and $P$. This is a contradiction. Consequently the assumption of the existence of $x^*$ is absurd.

The proof that (ii) implies case b) goes along the same way. $\qquad \square$

**Definition.** In the cases $b^\pm$) the orbits $x_1$ and $x_2$ are called **heteroclinic orbits** if $x_1 = x_2 \pmod 1$ one calls them **homoclinic orbits**. We denote the set of $x$, which are in case $b^\pm$) with $\mathcal{M}_{p/q}^\pm$.

---

**Theorem 2.6.2.** *If $x_1, x_2 \in \mathcal{M}(p/q)$ are neighboring, then there exist at least two non-periodic $x^+, x^- \in \mathcal{M}_{p/q}$, where $x^\pm$ is asymptotic to $x_2$ for $t \to \pm\infty$ and asymptotic to $x_1$ for $t \to \mp\infty$.*

---

*Proof.* Let $x_1(t)$ and $x_2(t)$ be two neighboring minimals in $\mathcal{M}(p/q)$. By Theorem 2.2.1 there exists for every $n \in \mathbb{N}$ a minimal $z_n(t)$ with $z_n(-n) = x_1(-n)$, $z_n(n) = x_2(n)$.

Call $x_m(t) = [x_1(t) + x_2(t)]/2$ the **middle line** of $x_1$ and $x_2$. By time translation, one can always achieve that

$$\tilde{z}_n(t) = z_n(t + \tau_n)$$

intersects the middle line $x_m$ in the interval $[0, q]$.

Because of the compactness proven in Theorem 2.4.9 there is a subsequence of $\tilde{z}_n$ which converges in $\mathcal{M}_{p/q}$ to an element $x^+$ which also intersects the middle line $x_m$ in the interval $[0, q]$. This $x_m$ is not periodic: between $x_1$ and $x_2$ there is by assumption no periodic minimal of type $(q, p)$.

It is obvious how one can construct $x^-$ analogously.                          □

**Example. Heteroclinic connection** of two neighboring geodesics (M. Morse 1924) [23].

We will see below that on the torus, two minimal neighboring closed geodesics of the same length can be connected by an asymptotic geodesic.

Theorems 2.6.1 and 2.6.2 can be summarized as follows:

---

**Theorem 2.6.3.** $\mathcal{M}_{p/q} = \mathcal{M}_{p/q}^+ \cup \mathcal{M}_{p/q}^- \cup \mathcal{M}(p/q)$. *If not* $\mathcal{M}_{p/q} = \mathcal{M}(p/q)$, *then* $\mathcal{M}_{p/q}^- \neq \emptyset$ *and* $\mathcal{M}_{p/q}^+ \neq \emptyset$.

---

**Appendix:** stability of periodic minimals.
A periodic extremal solution $x$ of type $(q, p)$ satisfies the Euler equation

$$\frac{d}{dt} F_p(t, x, \dot{x}) = F_x(t, x, \dot{x}) .$$

Let $\xi$ be a solution of the Jacobi equation

$$\frac{d}{dt}(F_{pp}\dot{\xi}) + (\frac{d}{dt}F_{px} - F_{xx})\xi = 0 .$$

We abbreviate this as

$$\frac{d}{dt}(a\dot{\xi}) + b\xi = 0, \ a = F_{pp}(t, x, \dot{x}) > 0 .$$

With $\xi(t)$ being a solution, also $\xi(t + q)$ is a solution and if $\xi_1$ and $\xi_2$ are two solutions, then the **Wronski determinant** $[\xi_1, \xi_2] := a(\dot{\xi}_1\xi_2 - \dot{\xi}_2\xi_1)$ is a constant.

It is different from zero if and only if $\xi_1$ and $\xi_2$ are linearly independent. In this case, there is a matrix $A$, so that

$$\begin{pmatrix} \xi_1(t+q) \\ \xi_2(t+q) \end{pmatrix} = A \begin{pmatrix} \xi_1(t) \\ \xi_2(t) \end{pmatrix} ,$$

or $W(t+q) = W(t)$ with $W = \begin{pmatrix} \xi_1 & \dot{\xi}_1 \\ \xi_2 & \dot{\xi}_2 \end{pmatrix}$. The comparison of the Wronskian

$$a(t+q)\det W(t+q) = [\xi_1, \xi_2](t+q) = [\xi_1, \xi_2](t) = a(t)\det W(t)$$

leads because of $a(t+q) = a(t) > 0$ to

$$\det(A) = 1 ,$$

and this means that with $\lambda$ also $\lambda^{-1}$ is an eigenvalue of $A$. There are three possibilities:

| | | |
|---|---|---|
| **Elliptic case** | $|\lambda| = 1, \lambda \neq \pm 1$ | (stable case ) |
| **Parabolic case** | $|\lambda| = \pm 1$ | |
| **Hyperbolic case** | $\lambda$ real, $\lambda \neq \pm 1$ | (unstable case) |

**Definition.** We say that the extremal solution $x$ is **elliptic**, **hyperbolic** or **parabolic**, if we are in the elliptic, the hyperbolic or the parabolic case.

It turns out that periodic minimals are not stable:

---

**Theorem 2.6.4.** *Periodic minimals* $\gamma \in \mathcal{M}(p/q), \gamma : t \mapsto x(t)$ *are not elliptic.*

---

*Proof.* We know that for all global minimals $\gamma \in \mathcal{M}(p/q)$ a solution $\xi \neq 0$ of the Jacobi equations has at most one root. If two roots would exist, there would be a conjugate point, which is excluded by Jacobi's Theorem 1.3.1. Assume now that $\gamma$ is elliptic. There is then by definition a complex solution $\zeta(t)$ of the Jacobi equation which satisfies

$$\zeta(t+q) = \lambda\zeta(t), |\lambda| = 1, \lambda = e^{i\alpha q} \neq 0, 1, \alpha \in \mathbb{R} .$$

For $\pi(t) = e^{-i\alpha t}\zeta(t)$ we have therefore

$$\pi(t+q) = \pi(t) .$$

Of course also

$$\xi(t) = \text{Re}\zeta(t) = \text{Re}(e^{i\alpha t}\pi(t))$$

is a solution of the Jacobi equation. From $e^{i\alpha q} \neq 0, 1$ follows that there exists $N > 1$ so that

$$\text{Re}(\exp(iN\alpha q)) < 0 .$$

This means that
$$\xi(t + Nq)\xi(t) < 0$$

so $\xi$ has a root $t \in [0, Nq]$. But also $t + kNq$ are roots for every $k \in \mathbb{N}$. This is a contradiction.                                                                                           $\square$

We show now that the situation is completely different for $n > 1$ and that the above argument does not apply. To do so, consider for $n = 2$ the integral

$$\int_{t_1}^{t_2} |\dot{x}^2 - \alpha Jx|^2 \, dt$$

with $x \in \text{Lip}(\mathbb{R}, \mathbb{R}^2)$, where $J = \begin{pmatrix} 0 & 1 \\ -1 & 0 \end{pmatrix}$ and where $\alpha$ is a real constant. In the class of periodic curves

$$x(t + 1) = x(t) \, ,$$

$x \equiv 0$ obviously is a minimal because

$$I(x)|_{t_1}^{t_2} = \int_{t_1}^{t_2} |\dot{x} - \alpha Jx|^2 \, dt \geq 0 \, .$$

On the other hand the Jacobi equation gives

$$\ddot{\xi} - 2\alpha J\dot{\xi} + \alpha^2 \xi = (\frac{d}{dt} - \alpha J)^2 \xi = 0 \, .$$

Let $c \in \mathbb{C}^2 \setminus \{0\}$ be a complex eigenvector of $J$, for example

$$c = \begin{pmatrix} 1 \\ i \end{pmatrix}, Jc = ic \, .$$

Obviously
$$\xi(t) = \text{Re}(e^{i\alpha t}c)$$

is a nontrivial solution of the Jacobi equation. This means that $x = 0$ is elliptic. However $x$ has no root. If $\xi(\tau) = 0$ were a root, we could achieve by translation that $\xi(0) = 0$ and so $\bar{c} = -c$. From

$$Jc = ic$$

would follow $Jc = -ic$ and $c = 0$. This would imply that $\xi$ is identical to 0.

This example shows also that for $n \geq 2$, periodic minimals can be elliptic.

**A remark on the average action:**

**Definition.** For $\gamma \in \mathcal{M}_\alpha$, define the **average action** as

$$\Phi(\gamma) = \lim_{T \to \infty} T^{-1} \int_0^T F(t, x, \dot{x}) \, dt \ .$$

---

**Theorem 2.6.5.** *a) For $\gamma \in \mathcal{M}_\alpha$ the average action is finite. It is independent of $\gamma$. We write therefore also $\Phi(\alpha) = \Phi(\gamma)$ with $\gamma \in \mathcal{M}_\alpha$.*
*b) On the set of rational numbers $\mathbb{Q}$, the map $\alpha \mapsto \Phi(\alpha)$ is strictly convex and Lipschitz continuous.*

---

We conjecture that $\alpha \mapsto \Phi(\alpha)$ is continuous on the whole real line $\mathbb{R}$.

*Proof.* a) For $\alpha = p/q$ and periodic $x$, the claim follows from

$$\Phi(\alpha) = q^{-1} \int_0^q F(t, x, \dot{x}) \, dt \ .$$

In the case $\alpha = p/q$, where $x$ is not periodic, the statement follows from the fact that $x$ is by Theorem 2.6.1 asymptotic to a periodic $\tilde{x}$.

For irrational $\alpha$ we can assume that $\gamma$ is in $\mathcal{M}_\alpha$, because non-recurrent orbits are asymptotic to recurrent orbits $\tilde{x} = u^{\pm}(t, \alpha t + \beta)$.

According to H. Weyl, there exists for every irrational $\alpha$ a Riemann integrable function $f(t, \theta)$ which is periodic in $t$ and $\theta$ so that

$$\lim_{T \to \infty} T^{-1} \int_0^T f(t, \alpha t + \beta) \, dt = \int_0^1 \int_0^1 f(t, \theta) \, dt d\theta \ .$$

One shows this first for $\exp(2\pi(kt + j\theta))$, then for trigonometric polynomials, then for continuous functions and finally, by lower approximation, for Riemann integrable functions. The claim follows if we put

$$f(t, \theta) = F(t, u^{\pm}(t, \theta), (\partial_t + \alpha \partial_\theta) u^{\pm}(t, \theta)) \ .$$

b) For $\alpha = p/q$, $\beta = p'/q'$, $\gamma = \rho\alpha + (1 - \rho)\beta$ with $\rho = s/r \in (0, 1)$ we get the inequality

$$\Phi(\gamma) < \rho\Phi(a) + (1 - \rho)\Phi(\beta) \ .$$

Let $x \in \mathcal{M}(p/q)$, $y \in \mathcal{M}(p'/q')$. If $x(t_0) = y(t_0)$ is the obligate intersection of $x$ and $y$ we define

$$z(t) = \begin{cases} x(t), & t \in [t_0, t_0 + qq's] \ , \\ y(t) - (p'q - pq')s, & t \in [t_0 + qq's, t_0 + qq'r] \ . \end{cases}$$

It is piecewise smooth, continuous and when continued periodically, $z$ has the rotation number

$$(p'q(r - s) + pq's)/(q'qr) = (1 - \rho)\beta + \rho\alpha = \gamma \, .$$

Because $z$ is not $C^2$ we have

$$\Phi(\gamma) < \frac{1}{qq'r} \int_0^{qq'r} F(t, z, \dot{z}) \, dt = \rho\Phi(\alpha) + (1 - \rho)\Phi(\beta) \, .$$

$\Phi$ is Lipschitz continuous because

$$
\begin{aligned}
\Phi(\gamma) - \Phi(\beta) \quad &< \quad \rho(\Phi(\alpha) - \Phi(\beta)) \\
&= \quad [(\gamma - \beta)/(\alpha - \beta)](\Phi(\alpha) - \Phi(\beta)) \\
&\leq \quad (\gamma - \beta)2 \max(\Phi(\alpha), \Phi(\beta))/(\alpha - \beta) \, . \qquad \square
\end{aligned}
$$

**Appendix: A degenerate variational problem on the torus.**

Finding $\mathcal{M}_\alpha$ for irrational $\alpha$ is computationally reduced to the determination of $u = u^\pm(t, \theta)$ because $u^+$ and $u^-$ agree almost everywhere. $u$ satisfied the equation (write $D$ for $\partial_t + \alpha\partial_\theta$)

$$DF_p(t, u, Du) = F_x(t, u, Du) \, .$$

These are the Euler equations to the variational problem

$$\int_0^1 \int_0^1 F(t, u, Du) \, dt d\theta \, ,$$

where $u(t, \theta) - \theta$ has period 1 in $t$ and $\theta$ and where $u(t, \theta)$ is monotone in $\theta$.

One could try to find $u$ directly. The difficulty with that is that for the minimum, whose existence one can prove, the validity of the Euler equation can not be verified so easily. It could be that the minimals are located at the boundary of the admissible functions. This can happen for example if $u$ is constant on an interval or if it has a point of discontinuity.

The problem can however be regularized if one looks at

$$\tilde{F}(t, \theta, u(t, \theta), \nabla u(t, \theta)) := \frac{\nu}{2} u_\theta^2 + F(t, u(t, \theta), Du(t, \theta))$$

with $\nu > 0$. One studies then the variational problem

$$\int_0^1 \int_0^1 \tilde{F}(t, u, \nabla u) \, dt \, d\theta$$

in the limit $\nu \to 0$ for $u(t, \theta) - \theta \in W^{1,2}(\mathbb{T}^2)$. It turns out that for $\nu > 0$ a minimal automatically is strictly monotone. This is done in [10].

## 2.7   Exercises to chapter II

1) Show that for a sequence $\gamma_n : t \mapsto x_n(t)$ in $\Xi$ one has $\gamma_n \to_w \gamma$ if and only if $x_n$ converges to $x$, the family $\{x_n\}$ is equicontinuous and if there exists $M \in \mathbb{R}$ so that $||\gamma_n||_\Xi \leq M$.

2) Prove the weak compactness of $K$ in the proof of Theorem 2.2.1 directly with the help of the theorem of Arzela–Ascoli.

3) Investigate the solutions of the nonlinear pendulum with $F = p^2/2 + (1/2\pi) \cdot \cos(2\pi x)$ and the corresponding Euler equations $\dot{x} = \sin(2\pi x)$ for minimality in the following cases:

a) A periodic oscillation $x(t) = x(t + T)$ with $x \neq 0$,
b) the stable equilibrium $x \equiv 0$,
c) the unstable equilibrium $x \equiv 1/2$.

4) Show that for $\gamma : t \mapsto x(t)$ with $\gamma \in \mathcal{M}$ the following holds: $\forall t_1, t_2 \in \mathbb{R}$

$$\frac{1}{t_2 - t_1} \int_{t_1}^{t_2} \dot{x}^2 \, dt \leq c \left\{ \left( \frac{x(t_2) - x(t_1)}{t_2 - t_1} \right)^2 + 1 \right\} .$$

# Chapter 3

# Discrete Systems, Applications

## 3.1 Monotone twist maps

In this chapter we consider situations which are closely related to the questions in Chapter II. Indeed, they are more or less the same questions, even though the assumptions are not identical. The topics require some small changes. But the underlying ideas remain the same.

The results of Mather apply to monotone twist maps, a topic which will appear now as an application of the earlier theory. Before we define these maps we derive them via a Poincaré map from the variational problem treated in Chapter II.

We assume that $F$ is given on the torus $\mathbb{T}^2$. We also assume that there are no extremal solutions in $[0, 1]$ which have conjugate points. This means that if $(t_1, x(t_1))$ and $(t_2, x(t_2))$ are conjugate points, then $t_2 - t_1 > 1$.

Under the assumptions of Chapter II, there exist solutions of the Euler equations

$$\frac{d}{dt} F_p = F_x$$

for all $t$. (See Exercise 1). Therefore the Poincaré map

$$f : S^1 \times \mathbb{R} \to S^1 \times \mathbb{R}, (x(0), \dot{x}(0)) \mapsto (x(1), \dot{x}(1))$$

is well defined on the cylinder $S^1 \times \mathbb{R} = \{t = 0, x \in S,$ $\dot{x} \in \mathbb{R}\,\}$, a hyper-surface in the phase space $\Omega \times \mathbb{R}$.

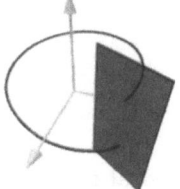

Let $x$ be a solution of the Euler equations. We define

$$x_0 := x(0), \ x_1 = x(1) \,,$$
$$y_0 := F_p(0, x_0, \dot{x}_0), \ y_1 := F_p(0, x_1, \dot{x}_1) \,,$$

and consider $x$ from now on as a function of $t, x_0$ and $x_1$. With

$$S(x_0, x_1) = \int_0^1 F(t, x, \dot{x}) \, dt$$

one has

$$S_{x_0} = \int_0^1 F_x \frac{dx}{dx_0} + F_p \frac{d\dot{x}}{dx_0} \, dt = \int_0^1 [F_x - \frac{d}{dt} F_p] \frac{dx}{dx_0} \, dt + F_p \frac{dx}{dx_0} |_0^1 = -y_0 \,,$$

$$S_{x_1} = \int_0^1 F_x \frac{dx}{dx_1} + F_p \frac{d\dot{x}}{dx_1} \, dt = \int_0^1 [F_x - \frac{d}{dt} F_p] \frac{dx}{dx_1} \, dt + F_p \frac{dx}{dx_1} |_0^1 = y_1 \,,$$

and (if $\dot{x}_0$ is considered as a function of $x_0$ and $x_1$),

$$S_{x_0 x_1} = -F_{pp}(0, x_0, \dot{x}_0) \frac{d\dot{x}_0}{dx_1} \,.$$

Because

$$\xi(t) := \frac{\partial x(t, x_0, x_1)}{\partial x_1}$$

is a solution of the Jacobi equation (differentiate $\partial_t F_p = F_x$ with respect to $x_1$) there are by assumption no conjugate points. Because $\xi(1) = 1$ and $\xi(0) = 0$ we have $\xi(t) > 0$ for $t \in (0,1)$ and this means

$$\dot{\xi}(0) = \frac{d\dot{x}_0}{dx_1} > 0$$

or $S_{x_0 x_1} < 0$. Summarizing, we can state that

$$f : (x_0, y_0) \mapsto (x_1, y_1)$$

satisfies

$$y_0 = -S_{x_0} \ , \quad y_1 = S_{x_1} \,,$$
$$S_{x_0 x_1} < 0 \ , \quad \text{which means} \ \frac{\partial y_1}{\partial x_0} > 0 \,.$$

In classical mechanics, $S$ is called a **generating function** for the canonical transformation $\phi$ (see [3]). The **Hamilton–Jacobi method** to integrate the Hamilton equations consists of finding a generating function $S$ in such a way that

$$H(t, x_0, S_{x_0}(x_0, x_1)) = K(x_1) \,.$$

The original Hamilton equations

$$\dot{x}_0 = H_{y_0}, \dot{y}_0 = -H_{x_0}$$

transform then to the integrable system

$$\dot{x}_1 = 0, \dot{y}_1 = K_{x_1} \ .$$

Many integrable systems in Hamiltonian mechanics can be solved with the Hamilton-Jacobi method. An example is the geodesic flow on the ellipsoid.

Instead of starting with the variational principle we could also define monotone twist maps directly:

**Definition.** A map

$$\phi : A \to A, (x, y) \mapsto (f(x, y), g(x, y)) = (x_1, y_1)$$

on the **annulus**

$$A = \{(x, y) \mid x \ (\text{mod } 1), a \leq y \leq b, \ -\infty \leq a < b \leq \infty \ \}$$

is called a **monotone twist map**, if it is an exact, boundary preserving $C^1$-diffeomorphism which has a continuation onto the cover $\tilde{A} = \mathbb{R} \times [a, b]$ of $A$:

- (0)  $f, g \in C^1(A)$ ,
- (i)  $f(x + 1, y) = f(x, y) + 1, \ g(x + 1, y) = g(x, y)$ ,
- (ii)  $a = ydx - y_1dx_1 = dh$ ,
- (iii)  $g(x, a) = a, \ g(x, b) = b$ ,
- (iv)  $\partial_y f(x, y) > 0$ .

In the cases when $a$ and $b$ are finite, one could replace the assumption (ii) also with the somehow weaker requirement of area-preservation:

$$dxdy = dx_1dy_1 \ .$$

The exact symplecticity (ii) follows from that. With the generating function $h$ from (ii), we can write these assumptions in a different but equivalent way. We write $h_i$ for the derivative of $h$ with respect to the $i$th variable.

- (0)'  $h \in C^2(\mathbb{R}^2)$ ,
- (i)'  $h(x + 1, x' + 1) = h(x, x')$ ,
- (ii)'  $y = -h_1(x, x_1), y_1 = h_2(x, x_1)$ ,
- (iii)'  $h_1(x, x') + h_2(x, x') = 0$, for $h_1(x, x') = a, b$ ,
- (iv)'  $h_{xx'} < 0$ .

We are interested in the orbits $(x_j, y_j) = \phi^j(x, y) \ (j \in \mathbb{Z})$ of the monotone twist map $\phi$. The dynamics given by $\phi$ is completely determined by the function $h$ which is defined on the torus $\mathbb{T}^2$ and which satisfies (0)' to (iv)'. The equations of motion

$$h_2(x_{j-1}, x_j) + h_1(x_j, x_{j+1}) = 0$$

form a second order difference equation on $\mathbb{T}^1$. It can be seen as the **Euler equations** to a variational principle.

It is not difficult to see that the function $h$ coincides with the generating function $S$ if $\phi$ is the Poincaré map.

The analogy between the continuous and the discrete case is as follows:

| continuous | | discrete | |
|---|---|---|---|
| $F(t, x, p)$ | Lagrange function | $h(x_j, x_{j+1})$ | generating function |
| $\int_{t_1}^{t_2} F(t, x, \dot{x})\, dt$ | action | $\sum_{j=n_1}^{n_2-1} h(x_j, x_{j+1})$ | action |
| $\frac{d}{dt} F_{\dot{x}} = F_x$ | Euler equation | $h_2 = -h_1$ | Euler equation |
| $F_{pp} > 0$ | Legendre condition | $h_{12} < 0$ | twist condition |
| $x(t)$ | extremal solution | $x_j$ | orbit |
| $x(t)$ | minimal | $x_j$ | minimal |
| $\dot{x}(t)$ | velocity | $\Delta x_j = x_{j+1} - x_j$ | first difference |
| $\ddot{x}(t)$ | acceleration | $x_{j+1} - 2x_j + x_{j-1}$ | second difference |
| $y = F_p(t, x, p)$ | momentum | $y_{j+1} = h_2(x_{j+1}, x_j)$ | momentum |

**Example. 1) The standard map of Taylor, Greene and Chirikov.**

Consider on the cylinder

$$\{(x, y) \mid x \pmod{1}, y \in \mathbb{R} \}$$

the map

$$\phi : \begin{pmatrix} x \\ y \end{pmatrix} \mapsto \begin{pmatrix} x + y + \frac{\lambda}{2\pi} \sin(2\pi x) \\ y + \frac{\lambda}{2\pi} \sin(2\pi x) \end{pmatrix} = \begin{pmatrix} x_1 \\ y_1 \end{pmatrix} .$$

Because

$$\begin{aligned}
\phi(x + 1, y) &= (x_1 + 1, y) = (x_1, y_1), \\
\phi(x, y + 1) &= (x_1 + 1, y + 1) = (x_1, y_1 + 1) ,
\end{aligned}$$

the map $\phi$ commutes with all elements of the fundamental group of the torus and can therefore be seen as a transformation on the torus. It has the generating function

$$h(x, x_1) = \frac{(x_1 - x)^2}{2} - \frac{\lambda}{2\pi} \cos(2\pi x) = \frac{(\Delta x)^2}{2} - \frac{\lambda}{2\pi} \cos(2\pi x) .$$

If one considers a few orbits of $\phi$, one often sees stable periodic orbits in the center the so-called 'stable islands'. The unstable, hyperbolic orbits are contained in a 'stochastic sea', which in the experiments typically appear as the closure of one orbit. Invariant curves which wind around the torus are called **KAM tori**. If the parameter value is increased — numerically one sees this for example at 0.97.. — then also the last KAM torus, the 'golden torus', vanishes. The name 'golden' originates from the fact that the rotation number is equal to the golden mean.

The formal analogy between discrete and continuous systems can be observed well with this example:

| continuous system | discrete system |
|---|---|
| $F(t, x, p) = \frac{p^2}{2} - \frac{\lambda}{4\pi^2} \cos(2\pi x)$ | $h(x_j, x_{j+1}) = \frac{(x_{j+1} - x_j)^2}{2} - \frac{\lambda}{4\pi^2} \cos(2\pi x_j)$ |
| $F_{pp} = 1 > 0$ | $h_{12} = -1 < 0$ |
| $\ddot{x} = \frac{\lambda}{2\pi} \sin(2\pi x)$ | $\Delta^2 x_j = \frac{\lambda}{2\pi} \sin(2\pi x_j)$ |
| $y = F_p = p$ | $y_{j+1} = h_2(x_j, x_{j+1}) = (x_{j+1} - x_j) = \Delta x_j$ |

There is an essential difference between the continuous system, the mathematical pendulum, and its discrete brother, the standard map. The continuous system is integrable: one can express $x(t)$ using elliptic integrals and Jacobi's elliptic function. The standard map however is not integrable for most parameter values. We will return to the standard map later.

**Example 2) Billiards.**

We take over the notation from the first section. The map

$$(s, t) \mapsto (s_1, t_1)$$

on the annulus $S^1 \times [0, \pi]$ becomes in the new coordinates

$$(x, y) = (s, -\cos(t)) ,$$

a map on the annulus $A = S \times [-1, 1]$. In order to show that

$$\phi : (x, y) \mapsto (x_1, y_1) = (f(x, y), g(x, y))$$

is a monotone twist map, we simply give a generating function

$$h(x, x_1) = -d(P, P_1) .$$

It has the properties $(0')$ until $(iv')$. Here $d(P, P_1)$ denotes the Euclidean distance between the points $P$ and $P_1$ on boundary of the table. These points are labeled by $x = s$ and $x_1 = s_1$ respectively.

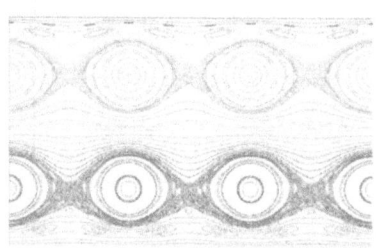    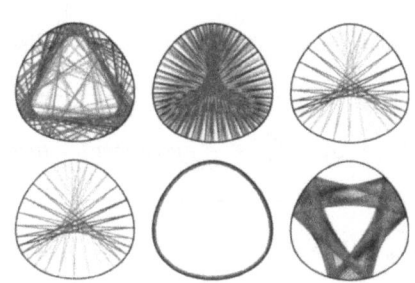

*Proof.* $(0')$ is satisfied if the curve is $C^2$.
$(i')$ is clear.
$(ii')$ $\cos(t) = h_x, -\cos(t_1) = h_{x_1}$.
$(iii')$ $x = f(x, 1)$ or $f(x, -1)$ implies $h_x + h_{x_1} = 0$.
$(iv')$ $h_{xx_1} < 0$ follows from the strict convexity of the curve.          □

**Example 3) Dual billiards.**

As in billiards, we start with a closed convex oriented  curve in the plane. We define a map $\phi$ on the exterior of $\Gamma$ as follows. From a given point $P \in E$ we draw the tangent $L$ to $\Gamma$ and denote by $Q$ the contact point which is at the center of the segment $L \cap \Gamma$. The line $L$ is chosen from the two possible tangents according to the orientation of $\Gamma$. We call $P_1$ the mirror of $P$ reflected at $Q$. The map $\phi$ which assigns to the point $P$ the point $P_1$ can be inverted. It is uniquely defined by the curve $\Gamma$. The emerged dynamical system is called **dual billiards**. The already posed questions, as for example the question of the existence of periodic points or the existence of invariant curves, appear here too.

There are additional problems which do not appear in billiards. One can for example ask for which $\Gamma$ every orbit is bounded or whether there are billiard tables $\Gamma$ for which there is an orbit which escapes to infinity. While this stability question is open in general, there is something known if $\Gamma$ is smooth or if it is a polygon as we will see later on.

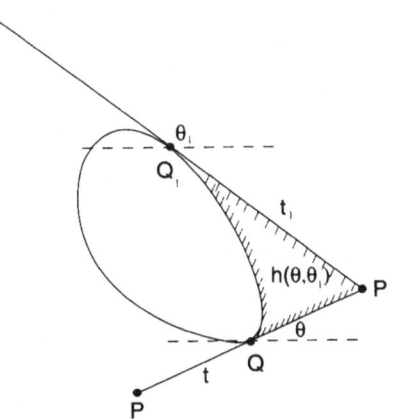

The dual billiards map $\phi$ has a generating function $h$. To find it, we use the coordinates

$$x = \theta/(2\pi),\ y = t^2/2\ ,$$

where $(t, \theta)$ are the polar coordinates of the vector $(P_1 - P)/2$. The generating function $h(x, x_1)$ is the area of the region between the line segments $QP_1, P_1Q_1$ and the curve segment of $\Gamma$ between $Q$ and $Q_1$. The map

$$\phi : (x, y) \mapsto (x_1, y_1)$$

is defined on the half-cylinder $A = S^1 \times [0, \infty)$ and the generating function $h$ satisfies properties $(0')$ until $(iv')$, if $\gamma$ is a convex $C^1$-curve. (Exercise).

**Periodic orbits.**

The existence of periodic orbits in monotone twist maps is guaranteed by the famous **fixed point theorem of Poincaré–Birkhoff** which we prove here only in a special case. In the next section we will look at the topic from the point of view of the theory developed in Chapter II and also will see that periodic orbits have to exist.

**Definition.** A map $\phi$,

$$(x, y) \mapsto (f(x, y), g(x, y)) = (x_1, y_1)$$

defined on the annulus $A = \{(x, y) \mid x \bmod 1, a \leq y \leq b, -\infty < a < b < +\infty\ \}$, is called a **twist map** if it has the following properties:
(0) $\phi$ is a homeomorphism of $A$.

(i) $f(x+1, y) = f(x, y) + 1, g(x+1, y) = g(x, y)$ (continuation onto a cover of $A$).
(ii) $dx dy = dx_1 dy_1$ (area preserving).
(iii) $g(x, y) = y$ for $y = a, b$ (preserving the boundary).
(iv) $f(x, a) - x > 0, f(x, b) - x < 0$ (twist map property).

---

**Theorem 3.1.1.** *(Poincaré–Birkhoff 1913) A twist map $\phi$ has at least two fixed points.*

---

A proof can be found in [6]. Unlike for monotone twist maps the composition of twist maps is again a twist map. As a corollary we obtain the existence of infinitely many periodic orbits:

---

**Corollary 3.1.2.** *For every twist map $\phi$ there is a $q_0$, so that for all $q > q_0$ there exist at least two periodic orbits of period $q$.*

---

*Proof.* Define

$$
\begin{aligned}
m &= \max\{f(x, a) - x \mid x \in \mathbb{R}\} < 0, \\
M &= \min\{f(x, b) - x \mid x \in \mathbb{R}\} > 0.
\end{aligned}
$$

We use the notation $\phi^j(x, y) = (f^j(x, y), g^j(x, y))$. For every $q > 0$, we have

$$
\begin{aligned}
\max(f^q(x, a) - x) &\leq \max\left\{ \sum_{j=0}^{q-1} f^{j+1}(x, a) - f^j(x, a) \right\} \leq qm < qM \\
&\leq \min\left\{ \sum_{j=0}^{q-1} f^{j+1}(x, b) - f^j(x, b) \right\} \leq \min\{f^q(x, b) - x\}.
\end{aligned}
$$

Let $q_0$ be so large that $q_0 M - q_0 m > 1$. If $q > q_0$, there is $p \in \mathbb{Z}$, such that $qm < p < qM$. And with

$$
\phi_{q,p} : (x, y) \mapsto (f^q(x, y) - p, g^q(x, y)),
$$

the twist maps satisfy

$$
\phi_{q,p}(x, a) < qm - p < 0 < qM - p < \phi_{q,p}(x, b).
$$

According to Poincaré–Birkhoff, the maps $\phi_{q,p}$ have at least two fixed points. This means that $\phi$ has two periodic orbits of type $(q, p)$. $\qquad\square$

It is easy to prove a special case of Theorem 3.1.1:

**Special case.**
A monotone twist map satisfying $f_y(x, y) > 0$ for all $(x, y) \in A$ has at least two fixed points, if the map rotates the boundaries of the annulus in opposite directions (property (iv) for twist maps).

*Proof.* Proof of the special case: because of the twist condition, there exists a function $z(x)$ satisfying

$$f(x, z(x)) = x .$$

The map $z$ is $C^1$ because of property (0) for the monotone twist maps. Thanks to area-preservation, the map must intersect the curve

$$\gamma : x \mapsto (x, z(x)) \in A$$

with the image curve $\phi(\gamma)$ in at least two points. These intersections define two fixed points of the map $\phi$. $\qquad\square$

**Definition.** By an **invariant curve** of a monotone twist map $\phi$ we mean a closed curve in the interior of $A$, which surrounds the inner boundary $\{y = a\,\}$ once and which is invariant under $\phi$.

From Birkhoff [12] is the following theorem:

---

**Theorem 3.1.3.** *(Birkhoff 1920) Every invariant curve of a monotone twist map is **star shaped**. This means that it has a representation as a graph $y = w(x)$ of a function $w$.*

---

For a careful proof see the appendix of Fathi in [15].

---

**Theorem 3.1.4.** *Every invariant curve of a monotone twist map can be represented as a graph $y = w(x)$ of a Lipschitz continuous function $w$.*

---

*Proof.* Let $\gamma$ be an invariant curve of the monotone twist map $\phi$. From Birkhoff's theorem we know that $\gamma$ is given as a graph of a function $w$. The map $\phi$ induced on $\gamma$ is a homeomorphism

$$(x, w(x)) \mapsto (\psi(x), w(\psi(x))) = (f(x, w(x)), g(x, w(x)))$$

given by a strictly monotone function $\psi$. Let $(x_j, y_j)$ and $(x'_j, y'_j)$ be two orbits on $\gamma$. Then $x_j$ and $x'_j$ are solutions of the Euler equations

$$
\begin{aligned}
-h_1(x_j, x_{j+1}) &= h_2(x_{j-1}, x_j) , \\
h_2(x'_{j-1}, x'_j) &= -h_1(x'_j, x'_{j+1}) .
\end{aligned}
$$

If we add both of these equations for $j = 0$ and add $h_1(x_0, x_1') - h_2(x_{-1}, x_0')$ on both sides, we get

$$h_2(x_{-1}', x_0') - h_2(x_{-1}, x_0') + h_1(x_0, x_1') - h_1(x_0, x_1)$$
$$= h_2(x_{-1}, x_0) - h_2(x_{-1}, x_0') + h_1(x_0, x_1') - h_1(x_0', x_1').$$

By the intermediate value theorem we have

$$\delta(x_{-1}' - x_{-1}) + \delta(x_1' - x_1) \le L(x_0' - x_0),$$

where $\delta = \min(-h_{12}) > 0$ and $L = \max(|h_{11}| + |h_{22}|) < \infty$. Because $x_1 = \psi(x_0), x_{-1} = \psi^{-1}(x_0)$, we have

$$|\psi(x_0') - \psi(x_0)|, |\psi^{-1}(x_0') - \psi^{-1}(x_0)| \le \frac{L}{\delta}|x_0' - x_0|.$$

This means that $\psi$ and $\psi^{-1}$ are Lipschitz continuous and also

$$w(x) = -h_1(x, \psi(x))$$

is Lipschitz continuous.                                                                                    $\square$

The question about the existence of invariant curves is closely related to stability:

**Definition.** The annulus $A$ is called a **region of instability**, if there is an orbit $(x_j, y_j)$ which goes from the inner boundary to the outer boundary. More precisely, this means that for all $\epsilon > 0$, there exists $n, m \in \mathbb{Z}$ so that

$$y_n \in U_\epsilon := \{a < y < a + \epsilon\},$$
$$y_m \in V_\epsilon := \{b - \epsilon < y < b\}.$$

---

**Theorem 3.1.5.** *$A$ is a region of instability if and only if there are no invariant curves in $A$.*

---

*Proof.* If there exists an invariant curve $\gamma$ in $A$, then this curve divides the annulus $A$ into two regions $A_a$ and $A_b$ in such a way that $A_a$ is bounded by $\gamma$ and the inner boundary $\{y = a\}$ and $A_b$ is bounded by the curves $\gamma$ and $\{y = b\}$. Because of the continuity of the map and the invariance of the boundary, the regions are mapped into themselves. $A$ can therefore not be a region of instability.

If $A$ is no region of instability, there exists $\epsilon > 0$, so that one orbit which starts in $U_\epsilon$ never reaches $V_\epsilon$. The $\phi$-invariant set

$$U = \bigcup_{j \in \mathbb{Z}} \phi^j(U)$$

is therefore disjoint from $V$. It is bounded by a $\phi$-invariant curve $\gamma$. According to Theorems 3.1.3 and 3.1.4 this curve is Lipschitz continuous.                              $\square$

One knows that for small perturbations of the **integrable monotone twist map**

$$\phi_\alpha : \begin{pmatrix} x \\ y \end{pmatrix} \mapsto \begin{pmatrix} x + \alpha(y) \\ y \end{pmatrix}, \quad \alpha'(y) \geq \delta > 0,$$

invariant curves with 'sufficiently irrational' rotation numbers survive. This is the statement of the **twist map theorem**, which is part of so-called KAM theory. See [24] for a reference to a proof.

**Definition.** The space $C^r(A)$ of $C^r$-diffeomorphisms on $A$ has the topology:

$$\|\phi_1 - \phi_2\|_r = \sup_{m+n \leq r} \left( \left| \frac{\partial^{m+n}(f_1 - f_2)}{\partial x^m \partial y^n} \right| + \left| \frac{\partial^{m+n}(g_1 - g_2)}{\partial x^m \partial y^n} \right| \right),$$

where $\phi_j(x, y) = (f_j(x, y), g_j(x, y))$.

**Definition.** We say that an irrational number $\beta$ is **Diophantine**, if there are positive constants $C$ and $\tau$, so that for all integers $p$ and $q > 0$ one has

$$|\beta - \frac{p}{q}| \geq Cq^{-\tau}.$$

---

**Theorem 3.1.6.** *(Twist map theorem)* *Given $\alpha \in C^r[a, b]$ with $r > 3$ and $\alpha'(y) \geq \delta > 0$, $\forall y \in [a, b]$, there exists $\epsilon > 0$, so that for every area-preserving $C^r$-diffeomorphism $\phi$ of $A$ with $\|\phi - \phi_\alpha\| < \epsilon$ and every Diophantine $\beta \in [\alpha(a), \alpha(b)]$, there exists an invariant $C^1$-curve $\gamma_\beta$. The map $\phi$ induces on $\gamma_\beta$ a $C^1$-diffeomorphism with rotation number $\beta$.*

---

**Remark.** For $r < 3$ there are counterexamples due to M. Hermann.

**Relating the continuous and the discrete systems:**

At the beginning of this section we have seen that if $F$ satisfies $F_{pp} > 0$ and is chosen so that no extremal solution has a conjugate point, then the Poincaré map $\phi$ has the generating function

$$h(x, x') = \int_0^1 F(t, x, \dot{x}) \, dt.$$

The map $\phi$ is then a monotone twist map. The exclusion of conjugate points was necessary. In general — if conjugate points are not excluded — one can represent the Poincaré map $\phi$ as a product of monotone twist maps: there exists $N \in \mathbb{N}$, so that the maps

$$\phi_{N,j} : (x(j/N), y(j/N)) \mapsto (x((j+1)/N), y((j+1)/N))$$

are monotone twist maps, if $(x(t), y(t))$ is a solution of the Hamilton equations

$$\dot{x} = H_y, \dot{y} = -H_x$$

and $H_{yy} > 0$. Each map

$$
\begin{aligned}
x(t + \epsilon) &= x(t) + \epsilon H_y + O(\epsilon^2) \,, \\
y(t + \epsilon) &= y(t) - \epsilon H_x + O(\epsilon^2) \,,
\end{aligned}
$$

is then a monotone twist map for small enough $\epsilon$. The Poincaré map $\phi$ can therefore be written as

$$\phi = \phi_{N,N-1} \circ \phi_{N,N-2} \circ \ldots \circ \phi_{N,0} \,.$$

We see that the extremal solutions of $\int F \, dt$ correspond to products of monotone twist maps.

The question now appears whether **every** monotone twist map can be obtained from a variational problem on the torus. For smooth $(C^\infty)$ maps, this is indeed the case ([25]). The result is:

---

**Theorem 3.1.7.** *(Interpolation theorem)   For every $C^\infty$ monotone twist map $\phi$ there is a Hamilton function $H = H(t, x, y) \in C^\infty(\mathbb{R} \times A)$ with*

$$
\begin{aligned}
&a) \qquad H(t + 1, x, y) = H(t, x, y) = H(t, x + 1, y) \,, \\
&b) \qquad H_x(t, x, y) = 0, y = a, b \,, \\
&c) \qquad H_{yy} > 0 \,,
\end{aligned}
$$

*so that the map $\phi$ agrees with $(x_0, y_0) \mapsto (x_1, y_1)$, where $(x(t), y(t))$ is a solution of*

$$\dot{x} = H_y(t, x, y), y = -H_x(t, x, y) \,.$$

---

With this interpolation theorem, Mather theory for monotone $C^\infty$ twist maps is a direct consequence of the theory developed in Chapter II.

## 3.2   A discrete variational problem

In this section we investigate a variational problem which is related to the problem treated in Chapter II. Rather than starting from the beginning, we just list the results which one can prove using the ideas developed in Chapter II. In [26] the proofs are made explicit for this situation. Let

$$\Phi = \{ x : \mathbb{Z} \mapsto \mathbb{R} \}$$

be the space of two-sided sequences of real numbers equipped with the product topology. An element $x \in \Phi$ is called a **trajectory** or an **orbit** and one can write $(x_j)_{j \in \mathbb{Z}}$ for $x$.

**Definition**. For a given function $h : R^2 \to R$ define

$$H(x_j, ..., x_k) = \sum_{i=j}^{k-1} h(x_i, x_{i+1})$$

and say, $(x_j, ..., x_k)$ is a **minimal segment** if

$$H(x_j + \xi_j, x_{j+1} + \xi_{j+1}, ..., x_k + \xi_k) \geq H(x_j, ..., x_k),$$

for all $\xi_j, ..., \xi_k \in \mathbb{R}$.

**Definition**. An orbit $(x_j)$ is called **minimal**, if every segment $(x_j, ..., x_k)$ is a minimal segment. One writes $\mathcal{M}$ for the set of minimal elements for $\Phi$. If $h \in C^2(\mathbb{R}^2)$, we say that $x$ is **stationary** or **extremal** if $x$ satisfies the Euler equations

$$h_2(x_{i-1}, x_i) + h_1(x_i, x_{i+1}) = 0, \forall i \in \mathbb{Z} .$$

Of course, every minimal orbit is extremal. We could ask that $h$ satisfies the conditions

(i) $\quad h(x, x') = h(x + 1, x' + 1)$ ,
(ii) $\quad h \in C^2(\mathbb{R})$ ,
(iii) $\quad h_{12}(x, y) \leq -\delta < 0$ .

The generating function of a monotone twist map satisfies these requirements. Additionally it also has the property

(iv) $\quad h_1(x, x') + h_2(x, x') = 0$, if $h_1(x, x') = a, b$ .

The theory can be developed also with fewer assumptions [4]: the requirements (ii) and (iii) can be replaced. Instead of (i) to (iii) it suffices to work with the following assumptions only:

(i') $\quad h(x, x') = h(x + 1, x' + 1)$
(ii') $\quad h \in C^2(\mathbb{R})$
(iii') $\quad h(x, x + \lambda) \to \infty$, uniform in x, $\lambda \to \infty$
(iv') $\quad x < x'$ or $y < y' \Rightarrow h(x, y) + h(x', y') < h(x, y') + h(x', y)$
(v') $\quad (x', x, x''), (y', x, y'')$ minimal $\Rightarrow (x' - y')(x'' - y'') < 0$.

Assumption (iii') follows from (iii) if $h \in C^2$ because of

$$-\lambda^2 \frac{\delta}{2} \geq \int_x^{x+\lambda} d\xi \int_\xi^{x+\lambda} h_{12}(\xi, \eta) \, d\eta$$

$$= -h(x, x + \lambda) + h(x + \lambda, x + \lambda) - \int_x^{x+\lambda} h_1(\xi, \xi) \, d\xi$$

$$= -h(x, x + \lambda) + O(\lambda) .$$

Assumption (iv$'$) is similar to (iii), because

$$-\delta(x'-x)(y'-y) \geq \int_x^{x'} \int_y^{y'} h_{12}(\xi,\eta) \ d\xi \ d\eta = h(x',y')+h(x,y)-h(x,y')-h(x',y)$$

and (v$'$) follows from (iii) by the monotonicity of $y \mapsto h_1(x,y)$ and $x \mapsto h_2(x,y)$. The assumption $x' < y'$ means $h_2(x',x) > h_2(y',x) > h_2(y',y)$ and $x'' < y''$ gives $h_1(x,x'') > h_1(x,y'') > h_1(y,y'')$. These inequalities together contradict the Euler equations $h_2(x',x) + h_1(x,x'') = 0$ and $h_2(y',y) + h_1(y,y'') = 0$.

We translate now the results and definitions in Chapter II to the current situation. The explicit translated proofs can be looked up in [4].

---

**Theorem 3.2.1.** *(Compare Theorem 2.4.1 or* [4], *3.16). For every* $(x_i)_{i\in\mathbb{Z}} \in \mathcal{M}$ *the rotation number* $\alpha = \lim_{i\to\infty} x_i/i$ *exists.*

---

For monotone twist maps, the rotation number is contained in the **twist interval** $[\alpha_a, \alpha_b]$, where $\alpha_a, \alpha_b$ are the rotation numbers of orbits which satisfy $h_1(x_j, x_{j+1}) = a$ (rsp. $h_1(x_j, x_{j+1}) = b$).

**Definition.** The **set of minimals with rotation number** $\alpha$ is denoted by $\mathcal{M}_\alpha$.

**Definition.** An orbit $x$ is called **periodic of type** $(q,p)$, if $x_{j+q} - p = x_j$. Call the set of these orbits $\mathcal{M}(q,p)$.

**Definition.** We say that two trajectories $(x_i)_{i\in\mathbb{Z}}$ and $(y_j)_{j\in\mathbb{Z}}$ **intersect**
a) **at the place** $k$, if $(x_{k-1} - y_{k-1})(x_{k+1} - y_{k+1}) < 0$ and $x_k = y_k$,
b) **between** $k$ **and** $k+1$, if $(x_k - y_k)(x_{k+1} - y_{k+1}) < 0$.

**Definition.** On $\mathcal{M}$ is defined the **partial order**

$$x \leq y \Leftrightarrow x_i \leq y_i, \forall i \in \mathbb{Z},$$

$$x < y \Leftrightarrow x_i < y_i, \forall i \in \mathbb{Z}.$$

The next result can be compared with Theorem 2.3.1 or [4], 3.1, 3.2, 3.9.

---

**Theorem 3.2.2.** *a) Two different minimal trajectories intersect at most once.*
*b) If* $x \leq y$, *then* $x = y$ *or* $x < y$.
*c) If* $\lim_{i\to\infty} |x_i - y_i| = 0$, *then* $x < y$ *or* $x > y$.
*d) Two different minimals of type* $(q,p)$ *do not intersect. The set* $\mathcal{M}(q,p)$ *is totally ordered.*

---

**Remark.** The strategy in the proof of Theorem 3.2.2 is the same as for Theorem 2.3.1. For a), we need the transversality condition (v$'$) as well as the order relation (iv$'$).

**Theorem 3.2.3.** *(Compare Theorem 2.3.3 or [4], 3.13). Minimals have no self in-tersections on* $\mathbb{T}^2$.

See Theorems 6.2 and 8.6 in [4] or [4] 3.3 and 3.17) as a comparison to the following theorem:

**Theorem 3.2.4.** *a) For every* $(q,p) \in \mathbb{Z}^2$ *with* $q \neq 0$, *there is a minimal of type* $(q,p)$.
*b)* $\mathcal{M}_\alpha \neq \emptyset$ *for all* $\alpha \in \mathbb{R}$.

For monotone twist maps this means that for every $\alpha$ in the twist interval $[\alpha_a, \alpha_b]$ there exist minimal trajectories with rotation number $\alpha$.

**Theorem 3.2.5.** *(Compare Theorem 2.5.9 or [4], 4.1). For irrational* $\alpha$ *the set* $\mathcal{M}_\alpha$ *is totally ordered.*

**Definition.** For $x \in \mathcal{M}_\alpha$ and irrational $\alpha$, define the maps $u^\pm : \mathbb{R} \mapsto \mathbb{R}$,

$$u^\pm : \alpha j - k \mapsto x_j - k$$

by closure of the two semicontinuous functions

$$u^+(\theta) = \lim_{\theta < \theta_n \to \theta} u(\theta_n)$$
$$u^-(\theta) = \lim_{\theta > \theta_n \to \theta} u(\theta_n) .$$

There are again two cases A) and B):
**case A):** $u^+ = u^- = u$
**case B):** $u^+ \neq u^-$.

**Theorem 3.2.6.** *(Compare Theorems 9.1, 9.13 or [4], 2.3.).* $u^\pm$ *are both strictly monotone in* $\theta$.

**Definition.** A trajectory $x \in \mathcal{M}_\alpha$ is called **recurrent**, if there exist $(j_m, k_m) \in \mathbb{Z}^2$, such that $x_{i+j_m} - k_m \to x_i$ for $m \to \infty$. The set of the recurrent trajectories is denoted by $\mathcal{M}^{rec}$. The elements of $\mathcal{M}_\alpha^{rec} = \mathcal{M}_\alpha \cap \mathcal{M}^{rec}$ are called **Mather sets** in case B). Define also

$$\mathcal{U}_\alpha := \{x \in \mathcal{M}_\alpha \mid x_j = u^+(\alpha j + \beta) \text{ or } x_j = u^-(\alpha j + \beta) \}$$

for $\beta \in \mathbb{R}$.

Compare the next result with Theorems 2.5.10–2.5.13 or 4.5, 4.6 in [4].

**Theorem 3.2.7.** *a)* $\mathcal{U}_\alpha = \mathcal{M}_\alpha^{rec}$.
*b)* $\mathcal{M}_\alpha$ *is independent of the element* $x$ *which generated* $u$.
*c)* $x \in \mathcal{M}_\alpha^{rec}$ *can be approximated by periodic minimals.*
*d) Every* $x \in \mathcal{M}_\alpha$ *is asymptotic to an element* $x^- \in \mathcal{M}_\alpha^{rec}$.

On $\mathcal{U}_\alpha^{rec} = \mathcal{M}_\alpha$ define the map

$$\psi : u(\theta) \mapsto u(\theta + \alpha) .$$

**Definition.** In the case when $h$ generates a monotone twist map $\phi$, we define for every irrational $\alpha \in [\alpha_a, \alpha_b]$ the set

$$\mathcal{M}_\alpha = \{(x, y) \mid x = u^\pm(\theta), \theta \in \mathbb{R}, y = -h_1(x, \psi(x)) \} .$$

**Theorem 3.2.8.** *(Mather, compare with 7.6 in [4]). If $h$ is a generating function for a monotone twist map on the annulus $A$, then for every irrational $\alpha$ in the twist interval $[\alpha_a, \alpha_b]$, one has:*
*a)* $\mathcal{M}_\alpha$ *is a non-empty subset of $A$, which is $\phi$-invariant.*
*b)* $\mathcal{M}_\alpha$ *is the graph of a Lipschitz function* $\omega : A_\alpha \to [a, b]$, *which is defined on the closed set* $A_\alpha = \{u^\pm(\theta) \mid \theta \in \mathbb{R}\}$ *by* $\omega(x) = -h_1(x, \psi(x))$.
*c) The map induced on* $\mathcal{M}_\alpha$ *is order-preserving.*
*d) The set* $A_\alpha$, *the projection of* $\mathcal{M}_\alpha$ *on* $S^1$ *is either the entire line* $\mathbb{R}$ *or it is a Cantor set. In the first case we are in case A) and the graph of* $\omega$ *is an invariant Lipschitz curve. In the second case we are in case B) and* $\mathcal{M}_\alpha$ *is called a* **Mather set** *with rotation number* $\alpha$.

We point to the recent papers of S.B. Angenent [2, 1], where these ideas are continued and generalized. In those papers, periodic orbits are constructed for monotone twist maps which do not need to be minimal but which have a prescribed index in the sense of **Morse theory**. In the proofs, **Conley's generalized Morse theory** is used. Furthermore, Angenent studied situations where the second order difference equations like $h_2(x_{i-1}, x_i) + h_1(x_i, x_{i+1}) = 0$ are replaced by higher order difference equations.

## 3.3   Three examples

In this section we return to the three examples of monotone twist maps which had been mentioned above: the standard map, billiards and the dual billiards.

**The standard map**

Mather has shown in [22] that the standard map has for parameter values $|\lambda| > 4/3$ no invariant curves in $A$. We show first, that for $|\lambda| > 2$, no invariant curves can exist.

According to Birkhoff's Theorem 3.1.4, an invariant curve is a graph of a Lipschitz function $y = \omega(x)$ on which the induced map is

$$x_1 = \psi(x) = f(x, \omega(x)) \,.$$

This map $\psi$ is a solution of the equation

$$h_1(x, \psi(x)) + h_2(\psi^{-1}(x), x) = 0 \,.$$

If we plug in

$$h_1(x, x_1) = -(x_1 - x) - \frac{\lambda}{2\pi} \sin(2\pi x) \,,$$
$$h_2(x, x_1) = x_1 - x_0 \,,$$

we get

$$-(\psi(x) - x) - \frac{\lambda}{2\pi} \sin(2\pi x) + x - \psi^{-1}(x) = 0$$

or

$$\psi(x) + \psi^{-1}(x) = 2x - \frac{\lambda}{2\pi} \sin(2\pi x) \,.$$

The left-hand side is a monotonically increasing Lipschitz continuous function. For $|\lambda| > 2$ we obtain a contradiction, because then the derivative on the right-hand side

$$2 - \lambda \cos(2\pi x)$$

has roots.

---

**Theorem 3.3.1.** *(Mather) The standard map has no invariant curves for parameter values $|\lambda| > 4/3$.*

---

*Proof.* We have even seen that the map $\psi$ which is induced on the invariant curve satisfies the equation

$$g(x) = \psi(x) + \psi^{-1}(x) = 2x - \frac{\lambda}{2\pi} \sin(2\pi x) \,.$$

For Lebesgue almost all $x$, we have

$$m := 2 - |\lambda| < g'(x) \leq 2 + |\lambda| =: M \,.$$

Denote by esssup$(f)$ the **essential supremum** of $f$ and by essinf$(f)$ the essential infimum. Let

$$R \;=\; \text{esssup } \psi'(x) \,,$$
$$r \;=\; \text{essinf } \psi'(x) \,.$$

Therefore, for almost all $x$,

$$r \;\leq\; \psi'(x) \leq R \,,$$
$$R^{-1} \;\leq\; (\psi^{-1})'(x) \leq r^{-1} \,,$$

and therefore

a)   $\max\{R + R^{-1}, r + r^{-1}\} \leq \max g'(x) \leq M$ ,

b)   $2 \min\{r, R^{-1}\} < r + R^{-1} \leq \min g'(x) = m$ .

From a) follows

$$\max(R, r^{-1}) \leq \frac{1}{2}(M + \sqrt{M^2 - 4}) \,.$$

From b) follows

$$\max(R, r^{-1}) \geq \min(R, r^{-1}) \geq \frac{2}{m} \,.$$

Together

$$\frac{2}{m} \leq \frac{1}{2}(M + \sqrt{M^2 - 4}) \,.$$

If we plug in $m = 2 - |\lambda|$ and $M = 2 + |\lambda|$, we obtain

$$(3|\lambda| - 4)|\lambda| \leq 0 \,.$$

Therefore, $|\lambda| \leq 4/3$.                                                                 $\square$

**Remarks**.

1) Theorem 3.3.1 was improved by Mac Kay and Percival in [19]. They could show the nonexistence of invariant curves for $|\lambda| > 63/64$.

2) Numerical experiments of Greene [13] suggest that at a critical value $\lambda = 0.971635...$, the last invariant curve disappears.

---

**Theorem 3.3.2.** *There exists $\epsilon > 0$ so that for $|\lambda| < \epsilon$ and for every Diophantine rotation number $\beta$, the set $\mathcal{M}_\beta$ is an invariant Lipschitz curve.*

---

*Proof.* Apply the twist Theorem 3.1.6. The function $\alpha(y)$ is of course given by $\alpha(y) = y$.                                                                 $\square$

**Remark.** Today there exist explicit bounds for $\epsilon$ [15]. Celletti and Chierchia have recently shown [8] that the standard map has analytic invariant curves for $|\lambda| \leq 0.65$.

A direct consequence of Theorem 3.2.7 and Theorem 3.3.1 is:

---

**Theorem 3.3.3.** *For every $\alpha \in \mathbb{R}$, there exist Mather sets $\mathcal{M}_\alpha$ for the standard map. For $\alpha = p/q$ there are periodic orbits of type $(q, p)$, for irrational $\alpha$ and $|\lambda| > 4/3$, the set $\mathcal{M}_\alpha$ projects onto a Cantor set.*

---

If we look at a few orbits of the standard map for different values of $\lambda$, the numerical calculations show the following picture:

In the unperturbed case $\lambda = 0$ all orbits are located on invariant curves. For $\lambda = 0.2$, the origin $(0, 0)$ is an elliptic fixed point. While increasing $\lambda$, for example for $\lambda = 0.4$, a region of instability grows near a hyperbolic fixed point. For $\lambda = 0.6$, there are still invariant KAM tori. For $\lambda = 0.8$ the dynamics is already quite complicated. For $\lambda = 1.0$ it is known that no invariant curves which wind around the torus can exist any more. For $\lambda = 1.2$, the 'stochastic sea' dominates already the regions of stability. One believes that for large $\lambda$, the dynamics is ergodic on a set of positive measure. For $\lambda = 10.0$ one can no more see islands even though their existence is not excluded.

### Birkhoff billiards

Also due to Mather [21] are examples of closed, smooth convex curves $\Gamma$ which define billiards with no invariant curves.

---

**Theorem 3.3.4.** *(Mather) If $\Gamma$ has a flat point, a point at which the curvature vanishes, then $\phi$ has no invariant curve.*

---

For example, the curve given by $x^4 + y^4 = 1$ has flat points.

*Proof.* If an invariant curve for the billiard map $\phi$ exists, then through every point $P$ of $\Gamma$ there would exist a minimal billiard trajectory. This means that it maximizes the length. We show that this can not be true for the flat point $P_0 \in \Gamma$.

If there would exist a minimal through $P_0$, we denote with $P_{-1}$ and $P_1$ the neighboring reflection points of the billiard orbit. We draw the ellipse, which passes through $P_0$ and which has both points $P_{-1}$ and $P_1$ as focal points. In a neighborhood of $P_0$, the curve $\Gamma$ is outside the ellipse, because $P_0$ is a flat point. This means that for a point $P \in \Gamma$ in a neighborhood of $P_0$, the length of the path $P_{-1}PP_1$ is bigger than the length of the path $P_{-1}P_0P_1$, which contradicts the minimality of the orbit (maximality of the length).

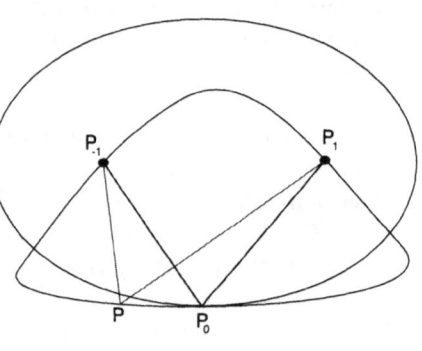

$\square$

**Definition.** A piecewise smooth closed curve $\gamma$ in the interior of the billiard table $\Gamma$ is called a **caustic** if the billiard orbit which is tangential to $\gamma$ stays tangent to $\Gamma$ after every reflection at $\gamma$.

A caustic of course leads to an invariant curve $\{(s, \psi(x))\}$ for the billiard map. In that case $\psi(s)$ is the initial angle of the billiard map path at the boundary which hits the caustic.

Lazutkin and Douady have proven [18, 11] that for a smooth billiard table $\Gamma$ with positive curvature everywhere, there always are "**whisper galleries**" near $\Gamma$.

---

**Theorem 3.3.5.** *If the curvature of the curve $\Gamma$ is positive everywhere and $\Gamma \in C^6$, there exist caustics near the curve $\Gamma$. These caustics correspond to invariant curves of the billiard map near $y = 0$ and $y = \pi$.*

---

From Hubacher [17] is the result that a discontinuity in the curvature of $\Gamma$ does not allow caustics near $\Gamma$.

---

**Theorem 3.3.6.** *If the curvature of $\Gamma$ has a discontinuity at a point, there exist no invariant curves in the annulus $A$ near $y = -1$ and $y = 1$.*

---

This theorem does not make statements about the global existence of invariant curves in the billiard map in this case. Indeed, there are examples where the curvature of $\Gamma$ has discontinuities, even though there are caustics.

A direct consequence of Theorem 3.2.7 is also the following result:

---

**Theorem 3.3.7.** *For every $\alpha \in (0,1)$, there are orbits of the billiard map with rotation number $\alpha$.*

---

**Appendix. Ergodic billiard of Bunimovich**.

**Definition**. An area preserving map $\phi$ of the annulus $A$ is called **ergodic**, if every $\phi$-invariant measurable subset of $\phi$ has Lebesgue measure 0 or 1.

If $\phi$ is ergodic, then $A$ is itself a region of instability. Moreover, there are then orbits in $A$, which come arbitrarily close to every point in $A$. This is called transitivity. Bunimovich [7] has given examples of ergodic billiards. Ergodic billiards have no invariant curves.

**Remark.** Mather theory still holds but not necessarily for the Bunimovich billiard, which produces a continuous but not a smooth billiard map.

**Dual Billards**

The dual billiard was suggested by B.H. Neumann (see [24]). Unlike in the case of billiards, affine equivalent curves produce affine equivalent orbits. Mathers Theorem 3.2.8 applied to this problem gives:

---

**Theorem 3.3.8.** *If $\Gamma$ is smooth, then there exists for every $\alpha \in (0,1)$ a point $(x, y)$ such that $(x, y) \notin \Gamma$ and such that the iterates rotate around $\gamma$ with an average angular speed $\alpha$.*

---

An application of the twist Theorem 3.1.6 with zero twist is the following theorem:

---

**Theorem 3.3.9.** *If the curve $\gamma$ is at least $C^r$ with $r > 4$, then every orbit of the dual billiards is bounded.*

---

Let $\Gamma$ be an arbitrary convex closed curve. For every angle $\psi \in [0, 2\pi)$ we construct the smallest strip bounded by two lines and which has slope $\arctan(\psi)$, and which contains the entire curve $\Gamma$. The two lines intersect $\Gamma$ in general in two intervals. Let $\xi$ be the vector which connects the center of the first interval with the center o the second interval. The convex closed curve $\gamma$ with polar representation $r(\psi) = |\xi_\psi|$ is called the **fundamental curve** of $\Gamma$.

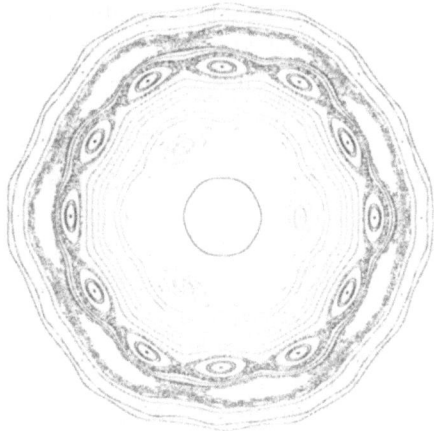

It is invariant under reflection at the
origin. The curve is therefore the
boundary of a unit ball in $\mathbb{R}^2$ with norm

$$||x|| = \min\{\lambda \in \mathbb{R} \mid \lambda x \in \gamma \} \ .$$

Denote by $\gamma^*$ the boundary of the unit
ball in the dual space of the Banach
space $(\mathbb{R}^2, || \cdot ||)$. This curve is called
the **dual fundamental curve of** $\Gamma$. Far
away from the curve $\Gamma$ the orbit is near
a curve which has the form of the dual
fundamental curve of $\Gamma$. If $\Gamma$ is a polygon, then also the dual fundamental
curve $\gamma^*$ of $\Gamma$ is a polygon. If the corners
of $\gamma$ have rational coordinates, then $\Gamma$
is called a **rational polygon**.

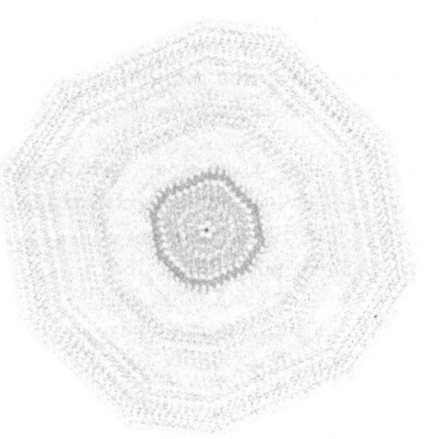

The following result is due to Vivaldi and Shaidenko [27] :

---

**Theorem 3.3.10.** *(Vivaldi and Shaidenko) If* $\Gamma$ *is a rational polygon, then all orbits
of the dual billiard are periodic. In this case there are invariant curves which are
close to the dual fundamental curve* $\gamma^*$ *of* $\Gamma$.

---

(Note added later: the proof in [27] had a gap but new proofs are available, see
Appendix).

**Open problem:** It is not known whether there exists a dual billiards for which
there are no invariant curves. In other words:

> **Problem.** Is it possible that for a convex curve $\gamma$ and a point
> $P$ outside of $\gamma$ the sequence $\phi_\gamma^n(P)$ is unbounded, where $\phi_\gamma$
> is the dual billiards map?

## 3.4   A second variational problem

Actually, one could find Mather sets in the discrete case by investigating a function $u$ satisfying the following properties:

(i)   $u$ is monotone.
(ii)  $u(\theta + 1) = u(\theta) + 1$.
(iii) $h_1(u(\theta), u(\theta + \alpha)) + h_2(u(\theta - \alpha), u(\theta)) = 0$ .

This is again a variational problem. Equation (iii) is the Euler equation describing extrema of the functional

$$I_\alpha(u) = \int_0^1 h(u(\theta), u(\theta + \alpha))\, d\theta$$

on the class $\mathcal{N}$ of the functions, which satisfy (i) and (ii). This is how Mather proved first the existence of $u^\pm$ [20]. A difficulty with this approach is to prove existence of the Euler equations. While this works formally:

$$
\begin{aligned}
\frac{d}{d\epsilon} I_\epsilon(u + \epsilon v)|_{\epsilon=0} &= \int_0^1 h_1(u, u(\theta + \alpha))v + h_2(u, u(\theta) + \alpha))v(\theta + \alpha)\, d\theta \\
&= \int_0^1 [h_1(u, u(\theta + \alpha)) + h_2(u(\theta - \alpha), u(\theta))]v(\theta)\, d\theta
\end{aligned}
$$

we can not vary arbitrarily in the class $\mathcal{N}$, because otherwise the monotonicity could get lost. Mather succeeded with a suitable parameterization.

A different possibility is to regularize the variational problem. Consider for every $\nu > 0$ the functional

$$I^{(\nu)}(u) = \int_0^1 \frac{\nu}{2} u_\theta^2 + h(u(\theta), u(\theta + \alpha))\, d\theta .$$

We look for a minimum in the class of functions $u$, for which $u(\theta) - \theta$ is a probability measure on $S^1$:

$$u - Id \in M^1(\mathbb{T}^1) .$$

The Euler equation to this problem is a differential-difference equation

$$-\nu u_{\theta\theta} + h_1(u(\theta), u(\theta + \alpha)) + h_2(u(\theta + \alpha), u(\theta)) = 0$$

for which one can show that the minimum $u_\theta^*$ is regular and monotone:

$$u_\nu^*(\theta) - \theta \in C^2(S^1),\ du_\nu^*(\theta)/d\theta > 0 .$$

Because the unit ball in $M^1(S^1)$ is weakly compact, the sequence $\nu_k \to 0$ has a subsequence $u_{\nu_k}^*$ which converges weakly to $u^*$ where $u^*$ satisfies the requirements (i) to (iii).

**Remark.** This strategy could maybe also be used to find Mather sets numerically.

## 3.5 Minimal geodesics on $\mathbb{T}^2$

Minimal geodesics on the torus were investigated already in 1932 by Hedlund [14]. In [5], Bangert has related and extended the results of Hedlund to the above theory. In this section, we describe this relation. For the proofs we refer to Bangert's article.

The two-dimensional torus $\mathbb{T}^2 = \mathbb{R}^2/\mathbb{Z}^2$ is equipped with a positive definite metric

$$ds^2 = g_{ij}(q)dq^i dq^j, g_{ij} \in C^2(\mathbb{T}^2) \ .$$

The length of a piecewise continuous curve $\gamma : [a,b] \to \mathbb{R}^2$ is measured with

$$L(\gamma) \;\; = \;\; \int_a^b F(q,\dot{q})\, dt \ ,$$

$$F(q,\dot{q}) \;\; = \;\; ([g_{ij}(q)\dot{q}^i \dot{q}^j])^{1/2} \ ,$$

and the distance of two points $p$ and $q$ is

$$d(q,p) = \inf\{L(\gamma) \mid \gamma(a) = p, \gamma(b) = q\} \ .$$

One calls such a metric a **Finsler metric**. A **Finsler metric** is a metric defined by $d$, where $F$ is homogeneous of degree 1 and satisfies the Legendre condition. The just defined metric generalizes the **Riemannian metric**, for which $g_{ij}$ is symmetric.

**Definition.** A curve $\gamma : \mathbb{R} \to \mathbb{R}^2$ called a **minimal geodesic** if for all $[a,b] \subset \mathbb{R}$ one has

$$d(\gamma(a), \gamma(b)) = L(\gamma)|_a^b \ .$$

Again we denote by $\mathcal{M}$ the set of minimal geodesics in $\mathbb{R}^2$.

Already in 1924, Morse had investigated minimal geodesics on covers of 2-dimensional Riemannian manifolds of genus $\geq 2$ [23]. Hedlund's result of 1934 was:

---

**Theorem 3.5.1.** *a) Two minimal geodesics intersect at most once.*
*b) There is a constant $D$, which only depends on $g$, so that every minimal geodesic is contained in a strip of width $2D$: $\exists$ constants $A, B, C$ with $A^2 + B^2 = 1$, so that for every minimal geodesic $\gamma : t \mapsto (q_1(t), q_2(t))$ one has*

$$|Aq_1(t) + Bq_2(t) + C| \leq D, \forall t \in \mathbb{R} \ .$$

*c) In every strip of this kind, there exists a geodesic: $\forall A, B, C$ with $A^2 + B^2 = 1$, $\exists$ minimal geodesic $\gamma : t \mapsto (q_1(t), q_2(t))$,*

$$|Aq_1(t) + Bq_2(t) + C| \leq D, \forall t \in \mathbb{R}$$

*with rotation number*

$$\alpha = -A/B = \lim_{t \to \infty} q_2(t)/q_1(t)$$

*which also can take the value $\infty$.*
*d) $\gamma \in \mathcal{M}$ has no self intersections on the torus.*
*e) If $\alpha$ is irrational, then $\mathcal{M}_\alpha$, the set of minimal geodesics with rotation number $\alpha$, is well ordered.*

---

How does this result relate to the theory developed in Chapter II? The variational problem which we had studied earlier is given by

$$I(\gamma) = \int_\gamma F(t, x, \dot{x}) \, dt = \int_\gamma F(q_1, q_2, \frac{dq_2}{dq_1}) \, dq_1 \;,$$

where $(t, x(t))$ is the graph of a function. Now we allow arbitrary curves $(q_1(t), q_2(t))$, which can in general not be written as graphs $q_2 = \phi(q_1)$. Also if we had $q_2 = \phi(q_1)$, as for example in the case of the Euclidean metric

$$F = [1 + (\frac{dq_2}{dq_1})^2]^{1/2} \;,$$

one has in general not quadratic growth. Bangert has shown how this problem can be avoided. We assume that the following existence theorem (compare [4], 6.1, 6.2) holds:

---

**Theorem 3.5.2.** *a) Two arbitrary points $p$ and $q$ on $\mathbb{R}^2$ can be connected by a minimal geodesic segment: $\exists \gamma^* : [a, b] \to \mathbb{R}^2, s \mapsto q^*(s)$ with $q^*(a) = p, q^*(b) = q$ and $L(\gamma^*) = d(p, q)$.*
*b) In every homotopy class $\{\gamma : s \mapsto q(s) \mid q(s + L) = q(s) + j, j \in \mathbb{Z}^2 \}$, there is at least one minimal. This minimal has no self intersections on $\mathbb{T}^2$.*

---

Let $\gamma : s \mapsto q(s) = (q_1(s), q_2(s))$ be a geodesic parametrized by the arc length $s$. According to the just stated theorem, there is a minimal $\gamma^* : s \mapsto q^*(s)$ with

$$q^*(s + L) = q^*(s) + e_2 \;,$$

where $e_2$ is the basis vector of the second coordinate. Because this minimal set has no self intersections, we can apply a coordinate transformation so that in the new coordinates

$$q_1(s) = 0, q_2(s) = s \;.$$

Therefore, one has

$$(k, s) = q^*(s) + k \;, \forall k \in \mathbb{Z} \;.$$

Define

$$h(\xi, \eta) := \overline{d}((0, \xi), (1, \eta)) \;,$$

where $\overline{d}$ is the metric $d$ in the new coordinate system. The length of a curve between $p$ and $q$ composed of minimal geodesic segments is given by

$$\sum_{j=1}^{r-1} h(x_j, x_{j+1}) \;.$$

The minimum

$$\min_{x_1=p}^{x_r=q} \sum_{j=1}^{r-1} h(x_j, x_{j+1})$$

is assumed by a minimal geodesic segment, which connects $(1, x_1)$ with $(r, x_2)$.

The following statement reduces the problem to the previously developed theory. It should be compared with 6.4 in [4].

---

**Theorem 3.5.3.** *The function h satisfies properties $(i')$ to $(iv')$.*

---

We can summarize the results as follows and compare them with [4], 6.5 up to 6.10:

---

**Theorem 3.5.4.** *a) For every $\alpha \in \mathbb{R}$, there exists a minimal geodesic with rotation number $\alpha$.*
*b) A minimal geodesic does not have self intersections on the torus.*
*c) Periodic minimal geodesics are minimal in their homotopy class.*
*d) Two different periodic minimal geodesics of the same period don't intersect.*
*e) A minimal geodesic $\gamma$ with rotation number $\alpha$ is either periodic or contained in a strip formed by two periodic minimal geodesics $\gamma^+$ and $\gamma^-$ of the same rotation number. In every time direction, $\gamma$ is asymptotic to exactly one geodesic $\gamma^+$ or $\gamma^-$. There are no further periodic minimal geodesics between $\gamma^+$ and $\gamma^-$. In other words, they are neighboring.*
*f) In every strip formed by two neighboring minimal periodic geodesics $\gamma^-$ and $\gamma^+$ of rotation number $\alpha$ there are heteroclinic connections in both directions.*
*g) Two different minimal geodesics with irrational rotation number do not intersect.*
*h) For irrational $\alpha$ there are two cases:*
**case A)**: *Through every point of $\mathbb{R}^2$ passes a recurrent minimal geodesic with rotation number $\alpha$.*
**case B)**: *The recurrent minimal geodesics of this rotation number intersect every minimal periodic geodesic in a Cantor set.*
*i) Every non-recurrent minimal geodesic of irrational rotation number $\alpha$ is enclosed by two minimal geodesics, which are asymptotic both forward and backwards.*
*j) Every non-recurrent minimal geodesic can be approximated by minimal geodesics.*

---

# 3.6 Hedlund's metric on $\mathbb{T}^3$

In this last section, we describe a metric on the three-dimensional torus as constructed by Hedlund. It shows that the above theory is restricted to dimension $n = 2$. The reason is that unlike in three dimensional space, non-parallel lines in $\mathbb{R}^2$ must intersect.

The main points are the following:

> 1) It is in general false that there exists for every direction a minimal in this direction. There are examples, where one has only three asymptotic directions.
>
> 2) It is in general false that if $\gamma^*(s+L) = \gamma^*(s) + k$ is minimal in this class $\mathcal{M}(L, k)$, then also $\gamma^*(s + NL) = \gamma^*(s) + NL$ is minimal in $\mathcal{M}(NL, Nk)$. Otherwise $\gamma^*$ would be a global minimal and would therefore be asymptotic to one of the three distinguished directions.

There are however at least $\dim(H_1(\mathbb{T}^3, \mathbb{R})) = 3$ minimals [5]:

---

**Theorem 3.6.1.** *On a compact manifold $M$ with $\dim(M) \geq 3$ and non-compact cover, there are at least $\dim(H_1(M, \mathbb{R})$ minimal geodesics.*

---

**The example:**

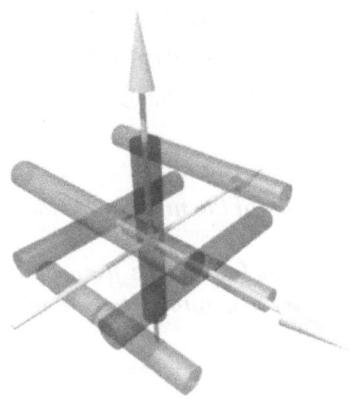

Define on the three dimensional torus $\mathbb{T}^3 = \mathbb{R}^3/\mathbb{Z}^3$ the metric

$$g_{ij}(x) = \eta^2(x)\delta_{ij} ,$$

where $\eta \in C^\infty(\mathbb{T}^3), \eta > 0$.

We need three closed curves $\gamma_1, \gamma_2$ and $\gamma_3$ on $\mathbb{T}^3$ which pairwise do not intersect.

Let $e_i$ denote the unit basis vectors in $\mathbb{R}^3$. Define

$$\gamma_1 : t \ \longmapsto \ te_1 \, ,$$

$$\gamma_2 : t \ \longmapsto \ te_2 + \frac{1}{2}te_1 \, ,$$

$$\gamma_3 : t \ \longmapsto \ te_3 + \frac{1}{2}te_2 + \frac{1}{2}te_1 \, ,$$

$$\Gamma \ = \ \bigcup_{j=1}^{3} \gamma_j \, .$$

We fix $0 < \epsilon < 10^{-2}$. The $\epsilon$-neighborhood $U_\epsilon(\gamma_i)$ form **thin channels** in $\mathbb{T}^3$, which do not intersect. Denote by

$$U(\gamma) = \bigcup_{j=1}^{3} U_\epsilon(\gamma_0)$$

the entire **channel system**.

Let $0 < \epsilon_i \le \epsilon < 10^{-2}$ for $i = 1, 2, 3$ and $\eta \in C^\infty(\mathbb{T}^3)$ with

$$i) \qquad \eta(x) \le 1 + \epsilon, \ \forall x \in \mathbb{T}^3 \, ,$$

$$ii) \qquad \eta(x) \ge 1, \ \forall x \in \mathbb{T}^3 \setminus U(\gamma) \, ,$$

$$iii) \qquad \eta(x) \ge \epsilon_i, \forall x \in U(\gamma_i) \setminus \gamma_i \, ,$$

$$iv) \qquad \eta(x) = \epsilon_i, \ \forall x \in \gamma_i \, .$$

The results are:

---

**Theorem 3.6.2.** *a) The total length of the minimal segments outside $U(\gamma)$ is smaller than 4.*
*b) Every minimal changes at most four times from one channel to another.*
*c) Every minimal is for $s \to \pm\infty$ asymptotic to one of the curves $\gamma_i$.*
*d) Every curve $\gamma_i$ is a minimal.*

---

*Proof.* We take first a piecewise $C^1$ curve,

$$\gamma : [a, b] \longmapsto \mathbb{R}^3, s \longmapsto \gamma(s)$$

parameterized by the arc length $s$. If $\eta^2 |\dot\gamma(s)|^2 = 1$, then

$$L(\gamma) = \int_a^b \eta|\dot\gamma| \, ds = \int_a^b \, ds = b - a \, .$$

Denote by $A$ the set of times, for which $\gamma$ is outside the channels

$$A = \{s \in [a, b] : \gamma(s) \notin U(\Gamma)\}$$

and let

$$\lambda(A) := \int_A ds \leq L(\gamma) \, .$$

Finally we need the vector $x = \gamma(b) - \gamma(a)$. To continue the proof we will need two lemmas. □

---

**Lemma 3.6.3.** *(Estimate of the time outside the channels). For every piecewise* $C^1$-*curve* $\gamma : [a, b] \to \mathbb{R}^3$, *we have*

$$\gamma(A) \leq \frac{11}{10}[L(\gamma) - \sum_{j=1}^3 \epsilon_j |x|_j] + 10^{-2} \, .$$

---

*Proof.* Define for $j = 1, 2, 3$,

$$A_j = \{s \in [a, b] \mid \gamma(s) \in U_\epsilon(\gamma_j)\} \, ,$$
$$A = \{s \in [a, b] \mid \gamma(s) \notin U(\Gamma)\} \, ,$$

so that $[a, b] = A \cup A_1 \cup A_2 \cup A_3$. If $n_j$ is the number of visits of $\gamma$ in $U_\epsilon(\gamma_i)$, then

$$\left| \int_{A_j} \dot\gamma_j \, ds \right| \leq 2n_j\epsilon, \quad i \neq j \, ,$$

$$\left| \int_{A_j} \dot\gamma_j \, ds \right| \geq \left| \int_{[a,b]} \dot\gamma_j \, ds \right| - \left| \int_{[a,b] \backslash A_j} \dot\gamma_j \, ds \right| = |x_j| - \left| \int_{[a,b] \backslash A_j} \dot\gamma_j \, ds \right| \, ,$$

$$\left| \int_{[a,b] \backslash A_j} \dot\gamma_j \, ds \right| \leq \int_A |\dot\gamma_j| \, ds + \sum_{i \neq j} \left| \int_{A_i} \dot\gamma_i \, ds \right| \geq \int_A \eta |\dot\gamma_j| \, ds + 2(n_i + n_k)\epsilon$$

$$= \lambda(A) + 2(n_i + n_k)\epsilon, \quad (\{i, j, k\} = \{1, 2, 3\}) \, .$$

We have

$$\lambda(A_j) = \int_{A_j} \eta |\dot\gamma| \, ds \geq \epsilon_j \left| \int_{A_j} \dot\gamma_j \, ds \right| \geq \epsilon_j \{|x_j| - \lambda(A) - 2(n_i + n_k)\epsilon\} \, .$$

Addition gives

$$\begin{aligned}
L(\gamma) &= \lambda(A) + \sum_{j=1}^3 \lambda(A_j) \\
&\geq \lambda(A) + (\sum_j \epsilon_j |x_j|) - 3\epsilon\lambda(A) - 4\epsilon^2(n_1 + n_2 + n_3) \\
&= (1 - 3\epsilon)\lambda(A) + \sum_{j=1}^3 [\epsilon_j |x_j| - 4\epsilon^2 n_j] \, .
\end{aligned}$$

On the other hand, there must be $n_1 + n_2 + n_3 - 1$ changes between channels and during these times the $\gamma$ are outside of $U_\epsilon(\Gamma)$. Because the distance between two channels is $\geq (1/2 - 2\epsilon)$,

$$\lambda(A) \geq \left( \sum_{j=1}^{3} n_j - 1 \right) (\frac{1}{2} - 2\epsilon)$$

follows. Therefore

$$\sum_{j=1}^{3} n_j \leq \lambda(A)(\frac{1}{2} - 2\epsilon)^{-1}$$

and

$$L(\gamma) \;\geq\; \lambda(A)[1 - 3\epsilon - 4\epsilon^2(\frac{1}{2} - 2\epsilon)^{-1} + \sum_{j=1}^{3} \epsilon_j |x_j| - 4\epsilon^2$$

$$\geq\; \frac{10}{11}(\lambda(A) + \sum_{j=1}^{3} \epsilon_j |x_j| - 4\epsilon^2) \; .$$

From this follows

$$\lambda(A) \leq \frac{11}{10}(L(\gamma) - \sum_{j=1}^{3} \epsilon_j |x_j|) + 10^{-2} \; . \qquad \square$$

---

**Lemma 3.6.4.**  *(Estimation of the length of a minimal).*

$$L(\gamma) = d(\gamma(a), \gamma(b)) \leq \sum_{j=1}^{3} \epsilon_j |x_j| + 3(1 + \epsilon) \; .$$

---

*Proof.* The length of a minimal from $\gamma(a)$ to one of the channels $U_\epsilon(\gamma_j)$ is less than or equal to $1 + \epsilon$. Also the length of a path which switches from $U_\epsilon(\gamma_j)$ to $U_\epsilon(\gamma_i)$ is smaller than or equal to $(1+\epsilon)$. The length of a path in a channel $U_\epsilon(\gamma_i)$ is smaller than $\epsilon_j |x_j|$. Therefore

$$L(\gamma) \leq 3(1 + \epsilon) + \sum_{j=1}^{3} \epsilon_j |x_j| \; . \qquad \square$$

*Continuation of the proof of Theorem* 3.6.2.
a) follows now directly from Lemma 3.6.3 and Lemma 3.6.4:

$$\lambda(A) \leq \frac{11}{10}(L(\gamma) - \sum_{j=1}^{3} \epsilon_j |x_j|) + 10^{-2} \leq \frac{11}{10}3(1 + \epsilon) + 10^{-2} < 4 \; .$$

b) Let $\gamma : [a, b] \to \mathbb{R}^2$ be a minimal segment, so that $\gamma(a)$ and $\gamma(b) \in U(\Gamma)$. We have

$$L(\gamma) \leq 2(1 + \epsilon) + 2\epsilon + \sum_{j=1}^{3} \epsilon_j |x_j| \, .$$

If $N$ is the number of times the channel is changed, then

$$N(\frac{1}{2} - 2\epsilon) \leq \lambda(A) \leq \frac{11}{10}[2(1 + \epsilon) + 2\epsilon + 10^{-2}]$$

which means $N < 5$ and therefore $N \leq 4$.

c) Because we only have finitely many changes a minimal $\gamma$ is finally contained in a channel $U_\epsilon(\gamma_k)$ and it is not difficult to see that $\gamma$ must be asymptotic to $\gamma_k$. (Exercise). $\qquad\square$

**Remark**. Again as an exercise it can be shown that for all $p, x \in \mathbb{R}^3$ one has

$$\sum_{i=1}^{3} \epsilon_i |x_i| - 4 \leq d(p, p + x) \leq \sum_{i=1}^{3} \epsilon_i |x_i| + 4$$

and with that we get the so-called **stable metric**

$$\tilde{d}(p, p + x) = \lim_{N \to \infty} \frac{d(p, p + Nx)}{N} = \sum_{j=1}^{3} \epsilon_j |x_j| \, .$$

The **stable norm** on $H_1(\mathbb{T}^3, \mathbb{R})$ is defined as follows. If $\gamma$ is a closed curve in $\mathbb{T}^3$ which represents an element in $H_1(\mathbb{T}^3, \mathbb{R})$, then the stable norm is defined as

$$||v|| = \tilde{d}(\gamma(0), \gamma(L)) \, .$$

It has a unit ball of the form of an octahedron. It turns out there is in general a close relation between the existence properties of minimal geodesics and the convexity of the unit ball in the stable norm. (See [5]).

## 3.7 Exercises to chapter III

1) Verify that for the billiard and for the dual billiard, the generating functions have properties $(0')$ through $(iv')$.

2) Show that in Hedlund's example, a minimal geodesic is always asymptotic to one of the curves $\gamma_k$.

3) Prove that the curves $\gamma_k$, $k = 1, 2, 3$ in Hedlund's example are minimal.

4) Verify in Hedlund's example the inequality

$$\sum_{i=1}^{3} \epsilon_i |x_i| - 4 \le d(p, p + x) \le \sum_{i=1}^{3} \epsilon_i |x_i| + 4 \ .$$

# Bibliography

[1] S. Angenent. Monotone recurrence relations their Birkhoff orbits and topological entropy. To appear in Ergodic theory and dynamical systems.

[2] S. Angenent. The periodic orbits of an area preserving twist map. *Commun. Math. Phys.*, 115:353–374, 1988.

[3] V.I. Arnold. *Mathematical Methods of classical mechanics.* Springer Verlag, New York, second edition, 1980.

[4] V. Bangert. Mather sets for twist maps and geodesics on tori. *Dynamics Reported*, 1:1–55, 1988.

[5] V. Bangert. Minimal geodesics. Preprint Mathematisches Institut Bern, 1988.

[6] M. Brown and W.D. Neuman. Proof of the Poincaré–Birkhoff fixed point theorem. *Mich. Math. J.*, 24:21–31, 1975.

[7] L.A. Bunimovich. On the ergodic properties of nowhere dispersing billiards. *Commun. Math. Phys.*, 65:295–312, 1979.

[8] A. Celletti and L.Chierchia. Construction of analytic KAM surfaces and effective stability bounds. *Commun. Math. Phys.*, 118:119–161, 1988.

[9] A.M. Davie. Singular minimizers in the calculus of variations. In *Proceedings of the International Congress of Mathematicians, Vol. 1, 2 (Berkeley, Calif., 1986)*, pages 900–905. AMS, Providence RI, 1987.

[10] J. Denzler. Mather sets for plane hamiltonian systems. *ZAMP*, 38:791–812, 1987.

[11] R. Douady. Application du théorème des tores invariantes. These 3 ème cycle, Université Paris VII, 1982.

[12] G.D.Birkhoff. Surface transformations and their dynamical applications. *Acta Math.*, 43:1–119, 1920.

[13] J. Greene. A method for determining a stochastic transition. *J. Math. Phys.*, 20:1183–1201, 1979.

[14] G.A. Hedlund. Geodesics on a two-dimensional riemannian manifold with periodic coefficients. *Annals of Mathematics*, 32:719–739, 1932.

[15] M.R. Herman. *Sur les courbes invariantes par les difféomorphismes de l'anneau. Vol. 1*, volume 103 of *Astérisque*. Société Mathématique de France, Paris, 1983.

[16] E. Hopf. Closed surfaces without conjugate points. *Proc. Nat. Acad. Sci. U.S.A.*, 34:47–51, 1948.

[17] A. Hubacher. Instability of the boundary in the billiard ball problem. *Commun. Math. Phys.*, 108:483–488, 1987.

[18] V.F. Lazutkin. The existence of caustics for a billiard problem in a convex domain. *Math. Izvestija*, 7:185–214, 1973.

[19] R.S. MacKay and I.C. Percival. Converse KAM: theory and practice. *Commun. Math. Phys.*, 98:469–512, 1985.

[20] J. Mather. Existence of quasi-periodic orbits for twist homeomorphism of the annulus. *Topology*, 21:457–467, 1982.

[21] J.N. Mather. Glancing billiards. *Ergod. Th. Dyn. Sys.*, 2:397–403, 1982.

[22] J.N. Mather. Nonexistence of invariant circles. *Ergod. Th. Dyn. Sys.*, 4:301–309, 1984.

[23] M. Morse. A fundamental class of geodesics on any closed surface of genus greater than one. *Trans. Am. Math. Soc.*, 26:25–65, 1924.

[24] J. Moser. *Stable and random Motion in dynamical systems*. Princeton University Press, Princeton, 1973.

[25] J. Moser. Break-down of stability. In *Nonlinear dynamics aspects of particle accelerators (Santa Margherita di Pula, 1985)*, volume 247 of *Lect. Notes in Phys.*, pages 492–518, 1986.

[26] J. Moser. On the construction of invariant curves and Mather sets via a regularized variational principle. In *Periodic Solutions of Hamiltonian Systems and Related Topics*, pages 221–234, 1987.

[27] F. Vivaldi and A. Shaidenko. Global stability of a class of discontinuous dual billiards. *Commun. Math. Phys.*, 110:625–640, 1987.

# Appendix A

# Remarks on the literature

> *Every problem in the calculus of variations has a solution, provided the word solution is suitably understood.*
>
> David Hilbert

Since these lectures were delivered by Moser, quite a bit of activity happened in this branch of dynamical system theory and calculus of variations. In this appendix some references to the literature are added. It goes without saying that this snapshot can not be exhaustive.

For the classical results in the calculus of variation see [36, 44]. In the meantime also the books [43, 91] have appeared. Notes of Hildebrandt [42] which were partly available in mimeographed form when the lectures had been delivered, have now entered the book [36]. This book is recommended to readers who want to know more about classical variational problems. Finally, one should also mention the review articles [73, 78].

More information about geodesic flows can be found in the sources [21, 13, 79]. Related to the theorem of Hopf are papers on integrable geodesic flows on the two-dimensional torus with Liouville metrics $g_{ij}(x, y) = (f(x) + h(y))\delta_{ij}$ (see [8, 69, 82]). For these metrics the flow has additionally to the energy integral $H(x, y, p, q) = (p^2 + q^2)/4(f(x) + h(y))$ also the quadratic integral $F(x, y, p, q) = (h(y)p - f(x)q)/4(f(x) + h(y))$. The problem to list all integrable geodesic flows on two dimensional Riemannian manifolds seems open (see [96]). Another theorem of Hopf type can be found in [80]. The higher dimensionsional generalization known under the name Hopf conjecture has been proven in [20].

More about Aubry–Mather theory can be found in [63]. Mather's first work is the paper [62]. The variational problem was later reformulated for invariant measures. It has been investigated further in [65, 67, 66, 68, 33]. See also the review [63].

Angenent's work mentioned in these lectures as preprints is published in [3]. The preprint of Bangert has appeared in [9].

The construction of Aubry–Mather sets as a closure of periodic minimals was done in [45, 46]. For a different approach to Aubry–Mather theory see [37]. While Golé's approach does not give all the results of Mather theory it has the advantage of being generalizable [51]. For higher dimensional Aubry–Mather theory, see [84]. For billiards, Aubry–Mather theory leads to average minimal action invariants [85]. The regularized variational principle mentioned in the course is described in detail in [75, 76]. For a reader who wants to learn more about the origins of the approach described in these notes, the papers [72, 29] are relevant. Aubry–Mather sets have been found as closed sets of weak solutions of the Hamilton–Jacobi equations $u_t + H(x, t, u)_x = 0$, which is a forced Burger equation $u_t + uu_x + V_x(x, t) = 0$ in the case $H(x, t, p) = p^2/2 + V(x, t)$ ([100]). Mané's work on Aubry–Mather theory announced in [60] appeared later in [61].

The theorem of Poincaré–Birkhoff which was first proven by Birkhoff in [14] has been given other proofs in [19, 64, 1].

For Aubry–Mather theory in higher dimensions, many questions are open. In [83] the average action was considered in higher dimensions. The higher-dimensional Frenkel–Kontorova model is treated in [84].

A good introduction to the theory of billiards is [92]. A careful proof for the existence of classes of periodic orbits in billiards can be found in [99]. A result analoguous to the theorem of Hopf for geodesics is proven in [12] or [102].

A question sometimes attributed to Birkhoff asks whether every smooth and strictly convex billiard is integrable. The problem is still open and also depends on the definition of integrability. Although Birkhoff made indications in [15, 16], he never seems have written down such a conjecture. The question was asked explicitly by H. Poritski in [81] who also started to attack the problem in that paper. The conjecture should therefore be called the Birkhoff–Poritski conjecture. For analytic entire perturbations, there is something known [28]. For more literature about caustics in billiards see [92, 49, 39].

More about the standard map can be found in the textbooks [86, 23, 47, 54]. The map appeared around 1960 in relation with the dynamics of electrons in microtrons [26]. It was first studied numerically by Taylor in 1968 and by Chirikov in 1969 (see [34, 25]). The map appears also by the name of the 'kicked rotator' and describes equilibrium states in the Frenkel–Kontorova model [52, 4].

The existence of stable 'islands' in the standard map for arbitrary large values of $\lambda$ has been proven by Duarte [30].

While it is known that for $\lambda \neq 0$, the standard map is non-integrable, has positive topological entropy and horse shoes (i.e., [31, 3, 35]), the question, whether hyperbolicity can hold on a set of positive Lebesgue measure stays open. While many area-preserving diffeomorphisms on the torus are known to be non-ergodic with positive topological entropy (i.e., [97, 104]), it is not known whether positive metric entropy is dense in the $C^\infty$ topology. The issue of the positivity of the Lyapunov exponents on some set of positive Lebesgue measure for Hamiltonian systems has been addressed at various places or reviews [70, 104, 25, 101, 55, 59, 41, 87, 90, 24, 27, 30, 103, 58, 98]. According to [30], the particular mathematical problem of positive entropy of the Chirikov standard map was promoted in the early 1980s by Sinai. The textbook [86] states a conjecture (H2) that the entropy of the Chirikov standard map is positive for all $\lambda > 0$ and that the entropy grows to infinity for $\lambda \to \infty$.

The break-up of invariant tori and the transition of "KAM Mather sets" to "Cantorous Mather sets" in particular has recently been an active research topic. The question, whether the MacKay fixed point exists is open. In a somewhat larger space of 'commuting pairs', the existence of a periodic orbit of period 3 was proven in [89]. A new approach to the question of the break-up of invariant curves is the theory of renormalisation in a space of Hamiltonian flows [50], where a nontrivial fixed point is conjectured also. For renormalisation approaches to the break-up of invariant curves one can consult [57, 88, 89, 50].

With the variational problem for twist maps one can also look for general critical points. An elegant construction of critical points is due to Aubry and Abramovici [7, 5, 6]. See [48] for a reformulation using the Percival functional. Aubry and Abramovici's approach shows that many Mather sets are hyperbolic sets. Hyperbolicity of Mather sets had first been demonstrated for the standard map in [38].

A part of the theory of the break-up of invariant curves was coined 'converse KAM theory' ([56]). Many papers appeared on this topic (i.e., [32, 11]).

The dual billiards is often also called 'exterior billiards' or 'Moser billiards'. The reason for the later name is that Moser often used it for illustrations, in papers or talks, for example in the paper [71] or in the book [70]. The question, whether a convex exterior billiards exists which has unbounded orbits is also open. Newer results on this dynamical system can be found in [93, 94, 18, 95]. Vivaldi

and Shaidenko's proof on the boundedness of rational exterior billiards had a gap.
A new proof has been given in [40] (see also [17]).

The different approaches to Mather theory are:

- Aubry's approach via minimal energy states. This was historically the first
  one and indicates connections with statistical mechanics and solid state
  physics.

- Mathers construction is a new piece of calculus of variation.

- Katok's construction via Birkhoff periodic orbits is maybe the technically
  most elegant proof.

- Golés proof leads to weaker results but has the advantage that it can be
  generalized.

- Bangert connected the theory to the classical calculus of variation and the
  theory of geodesics.

- Moser's viscosity proof is motivated by classical methods in the theory of
  partial differential equations.

Unlike for classical variational problems, where the aim is to find compact solutions
of differentiable functionals, the theme of these lectures shows that Mather theory
can be seen as a variational problem, where one looks for noncompact solutions
which are minimal with respect to compact perturbations. For such variational
problems the **existence** of solutions needs already quite a bit of work.

In an extended framework the subject leads to the theory of noncompact minimals, to the perturbation theory of non-compact pseudo-holomorphic curves on
tori with almost complex structure [77], to the theory of elliptic partial differential
equations [74, 22] or to the theory of minimal foliations [10].

As Hedlund's example shows, Mather's theory can not be extended to higher dimensions without modifications. The question arises for example what happens
with a minimal solution on an integrable three-dimensional torus if the metric is
deformed to the Hedlund metric. Another question is whether there is a Mather
theory which is applicable near the flat metric of the torus.

In [53] the Hedlund metric was investigated and the existence of many solutions
for the geodesic flow and non-integrability is proven. For metrics of the Hedlund
type on more general manifolds one can consult [2].

# Additional Bibliography

[1] S. Alpern and V.S. Prasad. Fixed points of area-preserving annulus homeomorphisms. In *Fixed point theory and applications (Marseille, 1989)*, pages 1–8. Longman Sci. Tech., Harlow, 1991.

[2] B. Ammann. Minimal geodesics and nilpotent fundamental groups. *Geom. Dedicata*, 67:129–148, 1997.

[3] S. Angenent. Monotone recurrence relations their Birkhoff orbits and topological entropy. *Ergod. Th. Dyn. Sys.*, 10:15–41, 1990.

[4] S. Aubry. Trajectories of the twist map with minimal action and connection with incommensurate structures. In *Common trends in particle and condensed matter physics (Les Houches, 1983)*, volume 103 of *Phys. Rep*, pages 127–141.

[5] S. Aubry. The concept of anti-integrability: definition, theorems and applications to the Standard map. In K.Meyer R.Mc Gehee, editor, *Twist mappings and their Applications*, IMA Volumes in Mathematics, Vol. 44. Springer Verlag, 1992.

[6] S. Aubry. The concept of anti-integrability: definition, theorems and applications to the Standard map. In *Twist mappings and their applications*, volume 44 of *IMA Vol. Math. Appl.*, pages 7–54. Springer, New York, 1992.

[7] S. Aubry and G.Abramovici. Chaotic trajectories in the Standard map. the concept of anti-integrability. *Physica D*, 43:199–219, 1990.

[8] I.K. Babenko and N.N. Nekhoroshev. Complex structures on two-dimensional tori that admit metrics with a nontrivial quadratic integral. *Mat. Zametki*, 58:643–652, 1995.

[9] V. Bangert. Minimal geodesics. *Ergod. Th. Dyn. Sys.*, 10:263–286, 1990.

[10] V. Bangert. Minimal foliations and laminations. In *Proceedings of the International Congress of Mathematicians, (Zürich, 1994)*, pages 453–464. Birkhäuser, Basel, 1995.

[11] U. Bessi. An analytic counterexample to the KAM theorem. *Ergod. Th. Dyn. Sys.*, 20:317–333, 2000.

[12] M. Bialy. Convex billiards and a theorem of E.Hopf. *Math. Z.*, 214:147–154, 1993.

[13] M.L. Bialy. Aubry-mather sets and Birkhoff's theorem for geodesic flows on the two-dimensional torus. *Commun. Math. Phys.*, 126:13–24, 1989.

[14] G. D. Birkhoff. An extension of Poincaré's last theorem. *Acta Math.*, 47:297–311, 1925.

[15] G.D. Birkhoff. On the periodic motions of dynamical systems. *Acta Math.*, 50:359–379, 1950.

[16] G.D. Birkhoff. *Dynamical systems*. Colloquium Publications, Vol. IX. American Mathematical Society, Providence, R.I., 1966.

[17] C. Blatter. Rationale duale Billards. *Elem. Math.*, 56:147–156, 2001.

[18] P. Boyland. Dual billiards, twist maps and impact oscillators. *Nonlinearity*, 9:1411–1438, 1996.

[19] M. Brown and W.D. Neuman. Proof of the Poincaré–Birkhoff fixed point theorem. *Mich. Math. J.*, 24:21–31, 1975.

[20] D. Burago and S. Ivanov. Riemannian tori without conjugate points are flat. *Geom. Funct. Anal.*, 4:259–269, 1994.

[21] M. Bialy (Byalyi) and L. Polterovich. Geodesic flows on the two-dimensional torus and phase transitions- commensurability-noncommensurability. *Funktsional. Anal. i Prilozhen*, 20:9–16, 1986. in Russian.

[22] L. A. Caffarelli and R. de la Llave. Planelike minimizers in periodic media. *Commun. Pure Appl. Math.*, 54:1403–1441, 2001.

[23] P. Le Calvez. Propriétés dynamiques des difféomorphismes de l'anneau et du tore. *Astérisque*, 204:1–131, 1991.

[24] L. Carleson. Stochastic models of some dynamical systems. In *Geometric aspects of functional analysis (1989-90)*, volume 1469 of *Lecture Notes in Math.*, pages 1–12. Springer, Berlin, 1991.

[25] B.V. Chirikov. A universal instability of many-dimensional oscillator systems. *Phys. Rep.*, 52:263–379, 1979.

[26] B.V. Chirikov. Particle confinement and adiabatic invariance. *Proc. R. Soc. Lond. A*, 413:145–156, 1987.

[27] R. de la Llave. Recent progress in classical mechanics. In *Mathematical Physics, X (Leipzig 1991)*, pages 3–19. Springer, Berlin, 1992. Also document Nr. 91-73 in *mp_arc@math.utexas.edu*.

[28] A. Delshams and R. Ramirez-Ros. On Birkhoff's conjecture about convex billiards. In *Proceedings of the 2nd Catalan Days on Applied Mathematics*. Presses Universitaires de Perpignan, 1995.

[29] J. Denzler. Studium globaler Minimaler eines Variationsproblems. Diplomarbeit ETH Zürich, WS 86/87.

[30] R. Duarte. Plenty of elliptic islands for the Standard family. *Ann. Inst. Henri Poincaré Phys. Théor.*, 11:359–409, 1994.

[31] E. Fontich. Transversal homoclinic points of a class of conservative diffeomorphisms. *J. Diff. Equ.*, 87:1–27, 1990.

[32] G. Forni. Analytic destruction of invariant circles. *Ergod. Th. Dyn. Sys.*, 14:267–298, 1994.

[33] G. Forni. Construction of invariant measures supported within the gaps of Aubry–Mather sets. *Ergod. Th. Dyn. Sys.*, 16:51–86, 1996.

[34] C. Froeschlé. A numerical study of the stochasticity of dynamical systems with two degree of freedom. *Astron. Astroph.*, 9:15–23, 1970.

[35] V.G. Gelfreich. A proof of the exponential small transversality of the separatrices for the Standard map. *Commun. Math. Phys.*, 201:155–216, 1999.

[36] M. Giaquinta and S. Hildebrandt. *Calculus of variations. I,II*, volume 310 of *Grundlehren der Mathematischen Wissenschaften*. Springer-Verlag, Berlin, 1996.

[37] C. Golé. A new proof of Aubry–Mather's theorem. *Math. Z.*, 210:441–448, 1992.

[38] D. Goroff. Hyperbolic sets for twist maps. *Ergod. Th. Dyn. Sys.*, 5:337–339, 1985.

[39] E. Gutkin and A. Katok. Caustics for inner and outer billiards. *Commun. Math. Phys.*, 173:101–133, 1995.

[40] E. Gutkin and N. Simányi. Dual polygonal billiards and necklace dynamics. *Commun. Math. Phys.*, 143:431–449, 1992.

[41] G.R. Hall. Some problems on dynamics of annulus maps. In *Hamiltonian dynamical systems (Boulder, CO, 1987)*, pages 135–152. Amer. Math. Soc., Providence, RI, 1988.

[42] S. Hildebrandt.   Variationsrechung und Hamilton'sche Mechanik.   Vor-
     lesungsskript Sommersemester 1977, Bonn, 1977.

[43] H. Hofer and E. Zehnder. *Symplectic invariants and Hamiltonian dynamics.*
     Birkhäuser Advanced Texts. Birkhäuser Verlag, Basel, 1994.

[44] J. Jost and X. Li-Jost. *Calculus of Variations*, volume 64 of *Cambridge
     studies in advanced mathematics*. Cambridge University Press, Cambridge,
     1998.

[45] A. Katok. Some remarks on Birkhoff and Mather twist map theorems. *Ergod.
     Th. Dyn. Sys.*, 2:185–194, 1982.

[46] A. Katok. Periodic and quasiperiodic orbits for twist maps. In *Dynamical
     systems and chaos (Sitges/Barcelona, 1982)*, pages 47–65. Springer, 1983.

[47] A. Katok and B. Hasselblatt. *Introduction to the modern theory of dynamical
     systems*, volume 54 of *Encyclopedia of Mathematics and its applications*.
     Cambridge University Press, 1995.

[48] O. Knill. Topological entropy of some Standard type monotone twist maps.
     *Trans. Am. Math. Soc.*, 348:2999–3013, 1996.

[49] O. Knill. On nonconvex caustics of convex billiards. *Elemente der Mathe-
     matik*, 53:89–106, 1998.

[50] H. Koch.  A renormalization group for hamiltonians, with applications to
     KAM tori. *Ergod. Th. Dyn. Sys.*, 19:475–521, 1999.

[51] H. Koch, R. de la Llave, and C. Radin. Aubry–Mather theory for functions
     on lattices. *Discrete Contin. Dynam. Systems*, 3:135–151, 1997.

[52] T. Kontorova and Y.I. Frenkel. *Zhurnal Eksper. i Teoret. Fiziki*, 8:1340–
     1349, 1938. Reference given in the Aubry–Le Daeron paper of 1983.

[53] M. Levi. Shadowing property of geodesics in Hedlund's metric. *Ergod. Th.
     Dyn. Sys.*, 17:187–203, 1997.

[54] A.J. Lichtenberg and M.A. Lieberman.  *Regular and Chaotic Dynamics*,
     volume 38 of *Applied Mathematical Sciences*. Springer Verlag, New York,
     second edition, 1992.

[55] R.S. MacKay. Transition to chaos for area-preserving maps. In *Nonlinear
     dynamics aspects of particle accelerators (Santa Margherita di Pula, 1985)*,
     volume 247 of *Lecture Notes in Phys.*, pages 390–454. Springer, Berlin-New
     York, 1986.

[56] R.S. MacKay. Converse KAM theory. In *Singular behavior and nonlinear dynamics, Vol. 1 (Smos, 1988)*, pages 109–113. World Sci. Publishing, Teaneck, NJ, 1989.

[57] R.S. MacKay. *Renormalisation in area-preserving maps*, volume 6 of *Advanced Series in Nonlinear Dynamics*. World Scientific Publishing Co., Inc., River Edge, NJ, 1993.

[58] R.S. MacKay. Recent progress and outstanding problems in Hamiltonian dynamics. *Physica D*, 86:122–133, 1995.

[59] R.S. MacKay and J.D. Meiss. Survey of Hamiltonian dynamics. In *Hamiltonian Dynamical Systems, a reprint selection*. Adam Hilger, Bristol and Philadelphia, 1987.

[60] R. Mané. Ergodic variational methods: new techniques and new problems. In *Proceedings of the International Congress of Mathematicians, Vol.1,2 (Zürich, 1994)*, pages 1216–1220. Birkhäuser, Basel, 1995.

[61] R. Mané. Lagrangian flows: the dynamics of globally minimizing orbits. *Bol. Soc. Brasil. Mat. (N.S.)*, 28:141–153, 1997.

[62] J. Mather. Existence of quasi-periodic orbits for twist homeomorphism of the annulus. *Topology*, 21:457–467, 1982.

[63] J. Mather and G. Forni. Action minimizing orbits in Hamiltonian systems. In *Transition to chaos in classical and quantum mechanics (Montecatini Terme, 1991)*, volume 1589 of *Lecture Notes in Math.*, pages 92–186. Springer, Berlin, 1994.

[64] J.N. Mather. Area preserving twist homeomorphism of the annulus. *Comment. Math. Helv.*, 54:397–404, 1979.

[65] J.N. Mather. Minimal measures. *Comment. Math. Helv.*, 64:375–394, 1989.

[66] J.N. Mather. Action minimizing invariant measures for positive definite Lagrangian systems. *Math. Z.*, 207:169–207, 1991.

[67] J.N. Mather. Variational construction of orbits of twist diffeomorphisms. *Journal of the AMS*, 4:207–263, 1991.

[68] J.N. Mather. Variational construction of connecting orbits. *Ann. Inst. Fourier (Grenoble)*, 43:1349–1386, 1993.

[69] V.S. Matveev. Square-integrable geodesic flows on the torus and the Klein bottle. *Regul. Khaoticheskaya Din.*, 2:96–102, 1997.

[70] J. Moser. *Stable and random Motion in dynamical systems*. Princeton University Press, Princeton, 1973.

[71] J. Moser. Is the solar system stable? *The Mathematical Intelligencer*, 1:65–71, 1978.

[72] J. Moser. Minimal solutions of variational problems on a torus. *Ann. Inst. Henri Poincaré Phys. Théor.*, 3:229–272, 1986.

[73] J. Moser. Recent developments in the theory of Hamiltonian systems. *SIAM Review*, 28:459–485, 1986.

[74] J. Moser. Quasi-periodic solutions of nonlinear elliptic partial differential equations. *Bol. Soc. Brasil. Mat. (N.S.)*, 20:29–45, 1989.

[75] J. Moser. Smooth approximation of Mather sets of monotone twist mappings. *Comm. Pure Appl. Math.*, 47:625–652, 1994.

[76] J. Moser. An unusual variational problem connected with Mather's theory for monotone twist mappings. In *Seminar on Dynamical Systems (St. Petersburg, 1991)*, pages 81–89. Birkhäuser, Basel, 1994.

[77] J. Moser. On the persistence of pseudo-holomorphic curves on an almost complex torus (with an appendix by Jürgen Pöschel). *Invent. Math.*, 119:401–442, 1995.

[78] J. Moser. Dynamical systems-past and present. In *Proceedings of the International Congress of Mathematicians, Vol. I (Berlin, 1998)*, pages 381–402, 1998.

[79] G.P. Paternain. *Geodesic Flows*. Birkhäuser, Boston, 1999.

[80] I.V. Polterovich. On a characterization of flat metrics on 2-torus. *J. Dynam. Control Systems*, 2:89–101, 1996.

[81] H. Poritsky. The billiard ball problem on a table with a convex boundary-an illustrative dynamical problem. *Annals of Mathematics*, 51:456–470, 1950.

[82] E.N. Selivanova and A.M. Stepin. On the dynamic properties of geodesic flows of Liouville metrics on a two-dimensional torus. *Tr. Mat. Inst. Steklova*, 216:158–175, 1997.

[83] W. Senn. Strikte Konvexität für Variationsprobleme auf dem n-dimensionalen Torus. *Manuscripta Math.*, 71:45–65, 1991.

[84] W.M. Senn. Phase-locking in the multidimensional Frenkel–Kontorova model. *Math. Z.*, 227:623–643, 1998.

[85] K-F. Siburg. Aubry–Mather theory and the inverse spectral problem for planar convex domains. *Israel J. Math.*, 113:285–304, 1999.

[86] Ya. G. Sinai. *Topics in ergodic theory*, volume 44 of *Princeton Mathematical Series*. Princeton University Press, Princeton, NJ, 1994.

[87] T. Spencer. Some rigorous results for random and quasiperiodic potentials. *Physica A*, 140:70–77, 1989.

[88] A. Stirnemann. Renormalization for golden circles. *Commun. Math. Phys.*, 152:369–431, 1993.

[89] A. Stirnemann. Towards an existence proof of MacKay's fixed point. *Commun. Math. Phys.*, 188:723–735, 1997.

[90] J.-M. Strelcyn. The coexistence problem for conservative dynamical systems: A review. *Colloq. Math.*, 62:331–345, 1991.

[91] M. Struwe. *Variational Methods*. Springer Verlag, 1990.

[92] S. Tabachnikov. *Billiards*. Panoramas et synthèses. Société Mathématique de France, 1995.

[93] S. Tabachnikov. On the dual billiard problem. *Adv. Math.*, 115:221–249, 1995.

[94] S. Tabachnikov. Asymptotic dynamics of the dual billiard transformation. *J. Stat. Phys.*, 83:27–37, 1996.

[95] S. Tabachnikov. Fagnano orbits of polygonal dual billiards. *Geom. Dedicata*, 77:279–286, 1999.

[96] I.A. Taĭmanov. Topology of Riemannian manifolds with integrable geodesic flows. *Trudy Mat. Inst. Steklov.*, 205:150–163, 1994.

[97] F. Takens. Homoclinic points in conservative systems. *Inv. Math.*, 18:267–292, 1972.

[98] M. Viana. Dynamics: a probabilistic and geometric perspective. *Doc. Math.J. DMV*, I:557–578, 1998. Extra Volume ICM, 1998.

[99] D.V. Treshchev V.V. Kozlov. *Billiards*, volume 89 of *Translations of mathematical monographs*. AMS, 1991.

[100] E. Weinan. Aubry–Mather theory and periodic solutions of the forced Burgers equation. *Commun. Pure Appl. Math.*, 52:811–828, 1999.

[101] A.S. Wightman. The mechanisms of stochasticity in classical dynamical systems. In *Perspectives in statistical physics*, Stud. Statist. Mech., IX, pages 343–363. North-Holland, Amsterdam-New York, 1981.

[102] M. Wojtkowski. Two applications of Jacobi fields to the billiard ball problem. *J. Diff. Geom.*, 40:155–164, 1994.

[103] L.-S. Young. Ergodic theory of differentiable dynamical systems. In *Real and complex dynamical systems (Hillerød, 1993)*, volume 464 of *NATO Adv. Sci. Inst. Ser. C Math. Phys. Sci.*, pages 293–336. Kluwer Acad. Publ., Dordrecht, 1995.

[104] E. Zehnder. Homoclinic points near elliptic fixed points. *Commun. Pure Appl. Math.*, 26:131–182, 1973.

# Index

# Advanced Courses in Mathematics CRM Barcelona

Since 1995 the Centre de Recerca Matemàtica (CRM) in Barcelona has conducted a number of annual Summer Schools at the post-doctoral or advanced graduate level. Sponsored mainly by the European Community, these Advanced Courses have usually been held at the CRM in Bellaterra.
The books in this series consist essentially of the expanded and embellished material presented by the authors in their lectures.